THE GREENIE

THE GREENIE

THE HISTORY OF WARFARE TECHNOLOGY
IN THE ROYAL NAVY

PATRICK A. MOORE

$PELLMOUNT

The Greenie is dedicated to those members of the Electrical Branch, its predecessors and successors, who gave their lives in the service of their country and whose memory was poignantly refreshed by the countless references to them and their extraordinary deeds encountered during the research for this book.

All royalties accrued for this title will be donated to Royal Navy charities under the auspices of the Collingwood Officers Association and the Royal Navy and Royal Marines Charities.

First published 2011 by
Spellmount, an imprint of
The History Press
The Mill, Brimscombe Port
Stroud, Gloucestershire, GL5 2QG
www.thehistorypress.co.uk

British Library Cataloguing in Publication Data.
A catalogue record for this book is available from the British Library.

ISBN 978 0 7524 6016 1

Typesetting and origination by The History Press
Printed in Great Britain

CONTENTS

ABBREVIATIONS

AB	Able Seaman
AB(EW)	Able Seaman (Electronic Warfare)
AB(M)	Able Seaman (Missile)
AB(R)	Able Seaman (Radar)
AB(S)	Able Seaman (Sonar)
AD	Action Data
ADA	Action Data Automation
ADAWS	Action Data Automated Weapons System
ADC	Action Data Communications
AE Branch	Air Engineering Branch
AFCT	Admiralty Fire Control Table
AFO	Admiralty Fleet Order
AIO	Action Information Organisation
AM	Amplitude Modulation
ARM	Availability, Reliability and Maintainability
ASW	Anti-submarine Warfare
AWT	Air Warfare Tactical
AWW	Air Warfare Weapons
BRNC	Britannia Royal Naval College
CACS	Computer Assisted Command System
CA(W)	Control Artificer (Weapons)
CCWEA	Charge Chief Weapons Engineering Artificer
CEA	Control, Electrical Artificer
CEM	Control Electrical Mechanic
CEW	Communications and Electronic Warfare
CDS	Comprehensive Display System
CINO	Chief Inspector Naval Ordnance
CIWS	Close In Weapons System
CNEO	Chief Naval Engineer Officer
COTS	Commercial Off The Shelf
CPOET	Chief Petty Officer Engineering Technician
DCI(RN)	Defence Council Instruction (Royal Navy)
DAMR	Department of Air Maintenance and Repair
DASW	Department of Anti-Submarine Warfare
DEE	Department of Electrical Engineering
DF	Direction Finding
DFCT	Dreyer Fire Control Table
DNAR	Department of Naval Air Radio
DNC	Directorate of Naval Construction
DNO	Director of Naval Ordnance

DNOA(E)	Director of Naval Officers Appointments (Engineering)
DRE	Department of Radio Equipment
DTM	Department of Torpedoes and Mines
EA	Electrical Artificer
EBD	Engineering Branch Development
EBWG	Engineering Branch Working Group
ECCM	Electronic Counter-Counter Measures
ECM	Electronic Counter Measures
EED	Department of Electrical Engineering
EM	Electrical Mechanic, or Electrician's Mate 1946–1955
ERA	Engine Room Artificer
ESM	Electronic Support Measures
ESO	Explosives Safety Officer
ET	Engineering Technician
EW	Electronic Warfare
FM	Frequency Modulation
GWS	Guided Weapons System
HACS	High Angle Control System
HF	High Frequency
IFF	Identification Friend or Foe
LEM	Leading Electrical Mechanic
LET	Leading Engineering Technician
LFLow Frequency	
LOM	Leading Operator Mechanic
LTO	Leading Torpedo Operator
LTO(LP)	Leading Torpedo Operator(Low Power)
MAD	Magnetic Anomaly Detection
ME Branch	Marine Engineering Branch
MEA(L)	Marine Engineering Artificer (Electrical)
MF	Medium Frequency
MEM(L)	Marine Engineering Mechanic (Electrical)
MEM(M)	Marine Engineering Mechanic (Mechanical)
MEO	Marine Engineering Officer
NDA	Naval Discipline Act
NLD	Naval Electrical Department
OA	Ordnance Artificer
OC	Ordnance Control
OEA	Ordnance Electrical Artificer
OEM	Ordnance Electrical Mechanic
OIC	Officer in Charge
OM(AW)	Operator Mechanic (Air Warfare)
OM(C)	Operator Mechanic (Communications)
OM(EW)	Operator Mechanic (Electronic Warfare)
OM(MW)	Operator Mechanic (Mine Warfare)
OM(UW)	Operator Mechanic (Underwater Warfare)
PEC	Printed Electronic Card
POET	Petty Officer Engineering Technician
PPI	Plan Position Indicator
PWO	Principal Warfare Officer
QO	Quarters Officer
Radar	Radio Detection and Ranging
RCM	Radar Counter Measures
RDF	Radio Direction Finding

REA	Radio Electrical Artificer
REM	Radio Electrical Mechanic
REME	Royal Electrical and Mechanical Engineers
RIS	Radio (or Radar) Interference Suppression
RNVR	Royal Navy Volunteer Reserve
Rx	Receiver
SCOT	Satellite Communications Operating Terminal
SD	Special Duties
SHF	Super High Frequency
SQ	Specialist Qualification
TAS	Torpedo and Anti-submarine
TBS	Talk Between ships
TGM	Torpedo Gunner's Mate
TM	Torpedo Mechanical
TS	Transmitting Station
Tx	Transmitter
UHF	Ultra High Frequency
VFO	Variable Frequency Oscillator
VHF	Very High Frequency
VST	Vent Sealing Tube
VT	Velocity Trigger or later Variable Time
WBD	Warfare Branch Development
WD	Weapons Data
WDO	Weapons Data and Ordnance
WEA	Weapons Engineering Artificer
WE Sub Branch	Weapons Engineering Sub Branch
WEE Branch	Weapons and Electrical Engineering Branch
WEM(O)	Weapons Engineering Mechanic (Ordnance)
WEM(C)	Weapons Engineering Mechanic (Control)
WEM(R)	Weapons Engineering Mechanic (Radio)
WEO	Weapons Engineering Officer
WRE Branch	Weapons and Radio Engineering Branch
WRNS	Women's Royal Naval Service
WT	Wireless Telegraphy
WTEE	Wireless Telegraphy Experimental Establishment
WM	Weapon Mechanician

ACKNOWLEDGEMENTS

The idea of documenting the story of the Electrical Branch arose during the run up to the 50th anniversary of the foundation of the Branch in 1996, and it became known as 'The Greenie Project'. The Editor of the *Review of Naval Engineering* magazine, at the time Lieutenant Commander Rod Chadwick, started the research work by using the magazine archives, going back to the 1940s, to document the early background of the Branch. Rod was assisted in his work by a team which comprised Lieutenant Pat Hunt, Lieutenant Mandy Clarke, Lieutenant Stewart Heather and Radio Supervisor Chris Rickard. This team extended the research using the Ministry of Defence archives in London and gathered anecdotal contributions, through interviews and letters, from many original members of the Electrical Branch and their successors. The resulting collection of material was incorporated into a first document that, regrettably, during the transformation of HMS *Collingwood* from a Weapon Engineering to a Maritime Warfare Establishment during the early 2000s, started to gather dust on a shelf. As the tide of change subsided, so the early work was brought to the attention of the committee of the Collingwood Officers' Association and they decided to back completion of the project. In 2007, the committee persuaded me to take on the task of writing and editing *The Greenie* and agreed that I should write the book such that it would be of interest not only to anyone who may have wielded a 'wee megger' in the Royal Navy but also the many civilians with association through either the Ministry of Defence or the defence industry.

During the book's development I felt obliged to reassess its scope as more information came to light on the earlier roots of electrical engineering in the Royal Navy and I discovered more about the close links between Branch restructuring and the major changes in naval warfare technology going back well into the nineteenth century. This further information, and the illustrative material used to eventually bring it to life, could not have been obtained without the help of the many organisations, both commercial and volunteer, that still exist in order to ensure the retention of the Royal Navy's engineering heritage. I would like to thank those organisations and the individuals who provided so much help in mining their respective archives for data and for supplying me with so many interesting images and giving me access to many documents which have remained undisclosed for well over 100 years. Those involved include the Naval Museum, Portsmouth (Stephen Courtney, Graham Muir), Naval Museum, Devonport (Jerry Rendle), Naval Historical Branch, Portsmouth (Jenny Wraight, Iain MacKenzie), BRNC Museum, Dartmouth (Richard Porter), Museum of Naval Firepower, Gosport (Derek Gurney), The Museum, HMS *Excellent* (Lieutenant Commander Brian Witts), HMS *Collingwood* (Commander Tim Stoneman, Keith Woodland), BAE Systems (Phil Stanton, Alison Gasser) and finally the Collingwood Radar and Communications Museum (Lieutenant Commander Bill Legg) where the full resources of the museum were made available for my use. In addition to the above sources, I used many others to correlate and contribute information in order to develop a coherent historical picture of what is an extensive subject. To all of the people who were so generous with their time and the authors of the many publications consulted and listed in the bibliography, I would like to extend my sincerest thanks. With those thanks comes a recommendation to anyone who finds *The Greenie* of interest to follow up with the organisations and books listed for a more in-depth coverage of the many subject areas covered in this book.

A special acknowledgement is due to the memory of 'JH' (Jack Hughes), 'Tugg', 'Sessions' and 'Dink', whose iconic cartoons for many years have lightened the reading of *Live Wire*, *Naval Electrical Review* and the *Admiralty Signals Establishment Bulletin*, all informal magazines used to communicate newsworthy engineering-related topics and feedback to and from the Fleet. My thanks to HMS *Collingwood* and the Collingwood Museum for allowing me the pleasure of resurrecting some of their work from these magazines. While the full publishing heritage of the works by these four cartoonists has proved to be elusive, it is hoped that use of their work to add colour and character to the story of *The Greenie* will be recognised as the tribute intended.

On a more general note, the writing of this book was made even more interesting by contributions from the many who thought that the history of the 'Greenie' was something worth recording. With an upfront apology for any transcription errors and editorial licence that may have been exercised on their original efforts made some time ago, I would like to thank all of those people whose personal memories of the Branch are chronicled in these pages. Unfortunately, space limitations have prevented the inclusion of all contributions passed to the original research team by 'Greenie' enthusiasts. For those contributions, I would like to express my grateful thanks and to apologise for the omission of anything that may have had special significance for the donor.

For their support in the actual production of *The Greenie*, I would like to record my thanks to David Riley for his restoration work on some of the early images and Lieutenant Commander (S) Richard Hart, and Lieutenant Commander (E)(WE) John Stafferton for their help in the proofreading of the text and constructive criticism of the content. Also, John's assistance with the daunting task of providing a comprehensive index, an essential element of a book such as this, was most welcome and certainly helped to lighten the editing load.

Finally, I would like to give thanks to the late Vice Admiral Sir Philip Watson KBE LVO CEng FIET CBIM for agreeing to preview my proofing version of *The Greenie* and the generous words he contributed to the Foreword. One of the most distinguished Greenies, Admiral Watson was a founder member of the Electrical Branch and he later became Director General Weapons and Chief Naval Engineering Officer. Prior to reaching those pinnacles of achievement for an electrical officer, he was also my Captain when I was appointed to *Collingwood* for Application Training in 1969. He might possibly have remembered this as we did spend every Monday morning for some three months giving our personal attention to a certain Leading Ordnance Electrical Mechanic at his table – me acting for the defence! It was a great sadness to me that this book did not appear until after Admiral Watson passed away on 9 December 2009.

Patrick A. Moore
Commander (E)(WE) Royal Navy

FOREWORD

I am delighted to write the Foreword to this outstanding work by Commander Moore and congratulate him on his historical review.

I was privileged to be appointed as the Assistant to Rear Admiral Bateson at the inauguration of the Electrical Branch on 1 January 1946.

The initial report by Rear Admiral Phillips, the subsequent report by Vice Admiral Gervaise Middleton and the Bateson Report were the result of the rapid expansion of electrical technology, throughout the Fleet, during the Second World War. Experience had been gained by commissioned officers, artificers and mechanics and there was a danger that this expertise would be lost. The Admiralty Board realised the importance of ensuring that these skills, gained in war, were not lost to the Royal Navy in peacetime.

The result of these early initiatives was the formation of the Electrical Branch on 1 January 1946. The Electrical Branch has enabled the Royal Navy to take advantage of the design, development and maintenance of electrical systems as technology has developed, so keeping the Royal Navy at the forefront of modern weapon systems.

I have pleasure in recommending this excellent book.

Vice Admiral Sir Philip Watson KBE LVO CEng FIET CBIM

INTRODUCTION

In the vernacular of the Royal Navy, the term 'Greenie' has evolved to describe the officers and ratings responsible for the support of electrical engineering functions in the ships of the Fleet. Although the reasons for the term can be loosely traced back to the introduction of the dark green distinction cloth stripe for warrant electricians in 1918, it really took off as a nom de guerre soon after 1946 when the unified Electrical Branch was first formed and all electrical officers wore the green stripe. It was at this point that the electrical engineering expertise of the Royal Navy became concentrated into one branch and one department on board ship, charged with the provision of all electrical engineering support to any other department in need of the expertise.

Although *The Greenie* is essentially the story of the electrical officers and men of the Royal Navy, it will quickly become apparent to the reader that other national service institutions also made significant contributions to the advancement of electrical technology in maritime warfare. These institutions include (in their past and present forms) the Royal Naval Volunteer Reserve, the Royal Naval Scientific Service and the Royal Naval Engineering Service. Many members of these services paid the ultimate sacrifice for their country in the pursuit of their work. It therefore gives me great pleasure to include at least some details of their efforts and achievements in the Greenie story.

While the name of the Electrical Branch was changed in 1961 to the Weapons and Radio Engineering Branch, and in 1965 to the Weapons and Electrical Engineering Branch, autonomy for electrical engineering expertise remained with a single branch until technology advances, particularly in machinery control systems, dictated the need for a more radical branch restructuring. This came in 1979 when, under the auspices of Engineering Branch Development, an Engineering Branch comprising Weapons Engineering, Marine Engineering and Air Engineering Sub Branches was formed. This change was primarily managed by transferring the requisite number of 'Greenies' (electrical officers, ordnance electrical artificers, mechanicians and mechanics) into the Marine Engineering Sub Branch.

Further advances in technology in the late 1980s resulted in Warfare Branch Development which involved the restructuring of the Operations Branch and Weapons Engineering Sub Branch in 1993 to form a Warfare Branch which embodied the concept of the user-maintainer. The impact on weapon engineering Greenies was that all rates and specialist categories of weapons engineering mechanic were transferred into the new Warfare Branch and merged with Operations Branch ratings. This combined pool of ratings was used to populate an interim rating structure, needed to support ship Schemes of Complement, which had been redesigned to allow for the combination of the operation and maintenance duties of each branch. At the same time, an operator mechanic rate was established and a training regime designed to support seagoing billets with both user and maintainer responsibilities was introduced.

In 2007, most of these user-maintainer changes were revoked, leading to redefined operator -only specialisations remaining in the Warfare Branch and the return of the Greenies into a streamlined Engineering Branch organised to meet the Royal Navy's needs for twenty-first century technical support. The streamlining involved the merging of the artificer and mechanic rates and all engineering ratings being given the new rate of 'engineering technician' but with specialist categories linked either to the weapon engineering or marine engineering functions as appropriate.

Despite this ebb and flow of electrical officers and ratings to the Marine Engineering and the Warfare Branches, the term 'Greenies' still lives on as a source of camaraderie, not for just those who originally formed the Electrical Branch but also the 'electrical specialists' who were embedded in those other branches. Perhaps more pointedly, it is still also used, but with more tongue in cheek endearment, by those retired and current members of the Royal Navy who still consider that the only real middle watch was the one spent on the ship's bridge!

While *The Greenie* explores the origins of Fleet manning structures, technology and the management of that technology at sea from medieval times until the present day, there is a particular focus on the significant increase in electrical technology introduced into the Fleet during the period 1890–1990 and the establishment of the Electrical Branch in 1946.

Against the backdrop of technological advances and structural volatility, this book recalls personal memories of Greenie Branch history and offers anecdotal reflections on the role of Branch personnel during times of hostility and momentous change. Finally, a number of traditional naval yarns and cameos with an Electrical Branch flavour have been included in order to provide light relief, should it be needed, during a story which, hopefully, will bring back fond memories to the many involved and raise the interest of those who were not.

CHAPTER 1

EARLY HISTORY

Technology, the application of scientific knowledge for practical purposes, can easily be thought of as a modern phenomenon. However, it has always been there in some guise or other and it is probably only the rate of change which has altered since the beginning of time. From the steady evolution of ideas, we have now reached the stage where the introduction of new technology is almost at a frenetic level. No area is free from its influence and the Royal Navy is no exception.

Prior to the nineteenth century, the Royal Navy's interest in technology was primarily based around the evolution of ship building and propulsion technology as applicable to the man of war, along with the specialist technology associated with the naval gun and navigation. It is these technological areas which helped form the initial manning structures aboard a warship, and some explanation of these generic beginnings is worth the reader's time in order to understand the social politics which had such an impact in the nineteenth and twentieth centuries on the way in which the Royal Navy was organised for the conduct of maritime warfare.

Ship Building

In broad terms, before the tenth century there were two types of seagoing ship evolving throughout the European continent, the galley and the sailing ship. Both types were being developed in a different fashion in two regions, northern Europe and the Mediterranean, with the latter examples evolving from as early as 5000BC beginning in Egypt. The way in which the two regions initially evolved their designs was very much dependent on the nature and availability of their land-based technology and the extent to which it could be transferred to the design and build of a ship. An example of this would be the carpentry techniques, which were more advanced and subtle in the Mediterranean region, and this showed in the way hulls were constructed.

Galley warfare. (Naval Historical Branch)

Another significant design driver was whether the ship was intended for military or commercial purposes with, arguably, early sailing designs being developed for commercial work and galley designs for the support of military activities.

The earliest naval warships were the galleys. A notable difference between the northern European and Mediterranean designs was that there is little evidence of the former being fitted with decking or a ramming bow, either under the waterline or above it. Thus, the inherent military capability of each design was different: the early northern European galley was more of a troop carrier, while the Mediterranean galley was better equipped to attack other shipping with central walkways and, in later designs, additional decks built over the oarsmen for the deployment of men specifically carried for fighting at sea. Given these two distinctive features, tactically the use of these two types of galley would have been very different.

Galleys could be sailed under certain wind conditions but the main method of propulsion was the oar and numerous oarsmen. In some cases, these oarsmen would have been slaves but in others, primarily in northern Europe, free men were employed. The use of oars did allow some manoeuvrability in inshore waters but, as a means of fighting at sea, the ships were limited in capability as they could not easily close to board each other without risk to manoeuvring. The exception to this was the use of the bow ram which provided an offensive, ship-sinking capability and acted as a bridge onto the opposing vessel for boarding purposes. However, unless decked over, the galley could not often carry sufficient additional numbers of fighting men and these men were only able to board on a narrow front to attack the enemy. Generally in northern Europe the oarsmen were free men and part of the fighting capability, so once they had put down their oars and picked up their weapons there were obvious tactical constraints. In the ninth century King Alfred was successful in using galleys in his defence of England against the Danes. He achieved this mainly by using tactics such as blocking the waterway exit once the enemy ships had been beached and, with land troops in support, effectively surrounding the shorebound raiding parties.

By the tenth century, oared galleys were falling out of use in northern Europe. The Bayeaux Tapestry shows the hull form used for the Norman invasion of 1066 to be similar to the early Norse designs but with complete reliance being placed on sail and no evidence of any rowing capability. However, the rowed galley continued to be developed for warfare use in the Mediterranean until as late as the eighteenth century. In fact, the Mediterranean galley is believed to have been the first ship to have any heavy ordnance fitted. This occurred around the fourteenth century when the 'Bombard' – a form of cannon which fired stone balls weighing some 50lbs – was introduced by the Venetians. Although the cannon gradually replaced the ram as the primary means of galley warfare at sea, it was often constrained by the presence of seated oarsmen to firing only fore and aft.

A Norman galley as depicted on the Bayeux Tapestry. (Naval Historical Branch)

Notwithstanding the occasional naval battle, the main reason for developing new ship designs was commerce. The fact that oarsmen took up space, had to be fed and, in the case of free men, paid, meant that the evolution of the galley was always going to languish because it was uneconomic for the transport of merchandise. The nature of the weather in northern Europe was also a factor against its development in that high seas were not the environment for using banks of oarsmen as the primary means of propulsion and the watertight integrity of the hull, with its numerous rowing ports, left much to be desired.

The first really effective craft for use as naval warships were single-masted ships which evolved from the type of sailing galley used by the Normans. This ship design was referred to as a 'Cog'. They were also known as 'Round Ships', so called because they often featured on the round civic seals of many of the English ports. This representation tended to distort the impression of the hull into a walnut shape and did not give an

The Civic Seal of Hastings, one of the Cinque Ports. (Hastings Museum)

impression of the great seaworthiness which was a feature of their reputation and the longevity of the basic design. In fact, the term 'cog' was probably a generic name which covered a number of similarly evolving northern European designs of the medieval sailing ship. The prime purpose of such a ship was commercial for they were designed to carry maximum amounts of cargo. However, they were also readily adaptable to the purpose of naval warfare.

What made the round ship suitable for military use were the high castle structures, usually fitted at the bow and stern of the ship, from which fighting men could launch missiles down onto the enemy once alongside at sea. The characteristics of castles, high freeboard and maximum use of hull space, meant that the ships were also ideal for high seas, commercial use when not co-opted into a military role. With the introduction of lightweight guns in the fifteenth century, this type of hull was the development baseline for the future ships of the line.[1]

Subsequent evolutionary drivers in design were the needs to increase the volume of the hull and speed through the water, both attributes being a requirement for longer range trading voyages and naval warfare expeditions. The conduct of these activities contributed further to ship design by giving more opportunities for the northern European and Mediterranean build features to cross-pollinate and evolve towards the most efficient and profitable design of sailing ship.

The *Great Harry*, an English carrack *c.*1545. (Naval Historical Branch)

One of these hybrid designs was the Carrack, developed in the fifteenth century by the Portuguese. Carracks were one of the first open ocean ships and they were used to explore the world by the Portuguese, Spanish and, in due course, the English. One notable example of the carrack was the *Victoria*, Magellan's ship that first completed the circumnavigation of the globe in 1522. The ships were large enough to be stable in heavy seas and had enough hull space to carry the provisions and trade goods needed for long exploration voyages. The high castles used in the design made the Carrack very defendable against smaller craft, a great asset in less welcoming parts of the world, and the stable deck gave an effective gun platform. The downside of a high castle was that the ship had a tendency to topple in high winds and manoeuvring was also affected. The Carrack was used as the basis for the design of royal ships produced for the English Navy until the middle of the sixteenth century, such as the *Henry Grace à Dieu,* or more colloquially 'The Great Harry'.

As more dedicated warships were built, two design requirements appeared which were not of great concern to the merchant fleet. The first was improved manoeuvrability and the second was greater hull strength to withstand the shocks of heavy gun operations and battle damage. These military requirements, plus the introduction of gun ports which allowed guns to be sited lower down in the ship, reduced the medieval high castle structures. This change improved windage across the bows and enabled the building of bigger ships with more guns capable of broadside firing. From the late sixteenth century, these fighting attributes were continuously developed and led to the iconic ships of the line that engendered the phrase 'The wooden walls of England' and dominated the naval warfare stage until the mid nineteenth century. The first HMS *Collingwood* provides a good example of the wooden hulled man of war at the peak of its development as pictured in an *Illustrated London News* article of 22 March 1845 (below).

In the mid nineteenth century, the next major change in ship building technology took place with the arrival of the ironclad hull. Iron hulled ships had been around since the late eighteenth century in the form of canal barges. Although the structural advantages were appealing, they were not immediately welcomed by the Royal Navy because of the problems with high seas navigation and action damage. The former became problematic because of induced magnetism in the iron hull distorting the earth's magnetic field and affecting the accuracy of the magnetic compasses in use. The latter was of concern because early trials had shown that iron plates of the time would shatter causing splinter hazards inside the ship or, in some cases, exit holes which were so ragged as to be impossible to plug against water ingress.

The problem of magnetic interference was overcome in the late 1830s when Sir George Airey developed a system of compensating iron balls that could be adjusted around the magnetic compass to counteract the induced magnetism of an individual ship. With this innovation, the Admiralty started to warm to the idea of iron hulls and, by 1843, there were seven small iron ships in Royal Navy service with further orders placed in 1845 for four frigates of around 3000 tons. Unfortunately, a report on a series of firing trials,[2] carried out against iron plating in 1846, again raised the profile of action damage concerns, such that the 1845 frigate programme was modified to convert the hulls to troop ships. However, the 1846 report did recommend that if the iron was backed with wood then splinter and exit damage would be reduced. In 1859, with this mitigation in mind, the Royal Navy started to build its first ironclad ship with

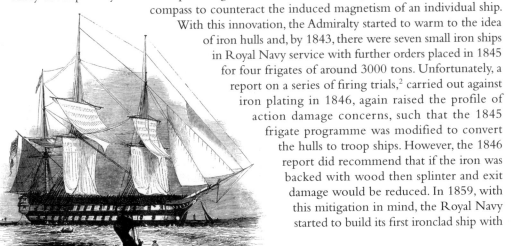

HMS *Collingwood, Illustrated London News,*
22 March 1845. (Collingwood Museum)

HMS *Warrior* in 1861.
(Naval Museum
Portsmouth)

a hull constructed using wrought iron plates backed by wood. The ship, HMS *Warrior,* completed build in 1861 and was fitted with a full sailing rig as well as a steam engine and propeller propulsion system.[3] The ironclad hull proved to be a short lived building technique as steel technology quickly advanced to provide a form of armour plating which allayed much of the concern about action damage. The introduction of steel hulls removed the need for timber, cut down the ship build time and enabled the construction of bigger and stronger ships, which could carry larger weapons and new systems based on emerging electrical technology. Meanwhile, *Warrior* was declared obsolete in 1883 without seeing any action or justifying her acquired reputation as the most powerful warship of her time.

Propulsion Systems

Although the use of oarsmen was a propulsion technique with its own place in the history of the galley, the main form of propulsion for the Royal Navy since its inception had been the sail.[4] The response of the Admiralty to the application of sailing technology was very much hands off and for centuries it was left for seagoing captains to develop and adapt new concepts of harnessing wind power for maritime warfare. Their Lordships' approach worked surprisingly well and many Captains were highly motivated – through the promise of prize money – to exploit the ship's mast arrangements with new sail combinations or experiment with the trim of their ships to maximise sailing performance in order to successfully pursue and defeat the enemy. This enthusiasm was also actively supported by crews anxious to supplement the not too generous pay of the Royal Navy.

There were few formal processes for putting common standards into place and the spread of new ideas was often based on word of mouth, letters and reports of proceedings which announced successful naval operations. It was not until the nineteenth century that any land-based form of sail training was put in place and, ironically, it was not too long afterwards that the Royal Navy's attention switched totally to steam propulsion.

The first steam-propelled ships were initially used by the Navy as tugs to assist sailing ships leave harbour in the face of head winds. One of the first of these was the naval steam vessel *Comet* which was built in 1822 and used as a steam training vessel before becoming a steam yacht for the Navy Board. In 1827, steamships started to be commissioned with the first warship, HMS *Dee,* appearing in 1829.

Apart from developing the use of the steam reciprocating engine as the prime motive force, the other technological debate at the time was whether the paddle wheel was more efficient than the propeller as a means of driving a ship through the water. Paddle wheel ships had been around since the late eighteenth century but the Royal Navy had only used them for support purposes as their seagoing performance, deck layout constraints and survivability were arguable for a fighting ship. By 1840, these open questions, along with rapid advances in screw propeller design from around

Parsons' Steam Turbine Engine destined for HMS *Vanguard*, 1907. (Naval Museum Portsmouth)

1800, had persuaded the Admiralty that the propeller was the way ahead and they started to place orders for a number of small screw-propeller-fitted ships in 1844. Notwithstanding this early confidence, largely for the benefit of the public, in 1845 it was decided to hold a series of trials using two ships, HMS *Alecto* and HMS *Rattler,* in a competition to establish which propulsion system was to be adopted for the Fleet. The two ships were of the same displacement and length and fitted with similar 200HP steam engines plus a sailing rig. The only significant difference was in the propulsion systems – *Alecto* had a side paddle wheel configuration while *Rattler* had a stern propeller fitted. The ships completed a series of time trials which involved the use of steam only, sail only and steam in combination with sail propulsion. The final, and most famous, trial engaged the two ships in a tug of war during which *Rattler* was able to pull *Alecto* astern through the water at some two knots despite her paddle wheels operating at full power. Throughout the trials, *Rattler* continued to demonstrate the superior efficiency of the propeller and, following this very public evaluation, the Royal Navy concentrated on a combination of the steam engine and the propeller, plus a sailing rig,[5] as its preferred form of propulsion system.

The next major advance in propulsion technology was the invention of the steam turbine, which eventually replaced the steam reciprocating engine. Although the principle of the turbine had been known for many years, the first modern steam turbine with marine propulsion applications was invented by George Parsons in 1884. As a ship propulsion system, it was initially fitted for demonstration in the *Turbinia* and appeared at the Diamond Jubilee Fleet Review in June 1997. The Parsons turbine was first fitted for the Royal Navy in HMS *Viper* in 1898, and subsequently it was used in HMS *Dreadnought* in 1906. Despite the advantages of economy shown by the steam turbine and the improved working conditions in the engine room, the steam reciprocating engine continued to be fitted in some Royal Navy ships until the end of the Second World War. The Loch Class was one of the last classes to be fitted with reciprocating engines and these ships saw service into the 1960s.

Naval Gunnery

In 1742, Benjamin Robins published *New Principles of Gunnery*, which introduced the science of ballistics, propellants and the effects on gun design. However, few of his ideas were implemented before the end of the eighteenth century and the naval gun continued to be developed around the concept of delivering an ever increasing weight of broadside to destroy the enemy. Poundage of shot was the British Fleet's measure of power projection, and even during Nelson's time the preferred option was to engage the enemy closely and fire as many broadsides at point blank range as possible.

It was often, once again, left to individual captains to innovate and take advantage of gun technology where they felt there were shortcomings in action. Robins' works on the physics and mathematics of gunnery firings were understood by only a few enthusiasts and the gun presented a different form of technical challenge to that of maintaining the sailing performance of the ship. While the former could often be conveniently ignored, albeit at the risk of failure in battle, the latter was always open for peer and public scrutiny to the benefit, or otherwise, of a captain's career.

It was the gunnery challenge that first generated a need for specialist onboard maintenance and training in the eighteenth century. The need for technological specialisation became more pressing with breech loading weapons, flintlock and electrical firing mechanisms, power operated training and elevating systems, centre line turrets and ammunition improvements being just a few significant examples of changes in gun technology introduced in the nineteenth century. Advances in gunnery and the application of technology in new warfare areas prompted the debate about how newly emerging electrical technology should be managed. One of the critical developments in gunnery was the exploding shell and its use at sea[6] as wooden hulls provided little defence against such ordnance. The result of this innovation was to accelerate the introduction of the armoured steel hull, which was not only capable of withstanding solid shot[7] but also offered the best defence against exploding shells. Later years would see the invention of the armour piercing shell, the response being to increase the thickness of armour. Thus battleship hull design started to evolve.

Emerging Electrical Technology

During the second half of the nineteenth century, the breadth and pace of technology advance throughout the Royal Navy started to increase rapidly. The Admiralty were obliged to cope with major changes such as the transition from sail to steam propulsion, and the move from wood to ironclad and then armoured hulls. The supremacy of the Royal Navy was being challenged by other growing navies' use of technology, and any notion that technological advances could be ignored was quickly dismissed by the threats being perceived from those nations who sought to test the resolve of the Grand Fleet.

Within the dynamic technical environment of the nineteenth century the newly discovered uses of electricity were soon applied at sea. Following the initial introduction of the electrically

HMS *Excellent* outrigger torpedo trials, Gig *c.*1860. (Naval Historical Branch)

detonated outrigger torpedo, other ship-borne applications for electrical technology swiftly began to emerge and these were soon to have an influence on the future development of the Royal Navy far greater than originally could have been predicted. For brevity, the term 'electrical technology' in this book refers to all forms of the engineering application of the phenomenon of electricity, including electronic, solid state, microchip and software technology.

Social Change

In parallel with technology changes, the nineteenth century also saw many social changes taking place. With the emergence of steam technology, naval officers, well versed in the art of fighting a sailing ship and usually possessing the same seamanship skills as their crew, found that their hard-won knowledge was no longer essential for conducting naval warfare as mechanical propulsion systems began replacing sail. With the arrival of the steam engine and the engineer, many of the traditional divisions between officer and rating started to be challenged. It was no coincidence that this occurred in parallel with the improvement in general educational standards. The lower classes, an historical source of rating recruits, seized the opportunity of the new technology to advance themselves. This energy fuelled the engine of the Victorian industrial revolution. With such newfound technical expertise, what had been considered the lower classes evolved social and professional aspirations that the Admiralty was to find an unstoppable force for change in the service.

As far as manning was concerned, the feudal system of the medieval era had been replaced in the eighteenth century by the 'Hire and Fire' approach to recruiting, which relied on the press gang and appeals to Jolly Jack's sense of duty. In the nineteenth century it would have to change yet again and become more inclusive because of the technological needs of the Royal Navy which, to an extent, were anathema to many in the existing officer class. This new social enlightenment led to many questions about the Royal Navy as an employer and the answers raised demands for such things as improved living conditions, defined conditions of service, the supply of free uniforms, better rates of pay and more inclusive pensions in order to stop crews voting with their feet.

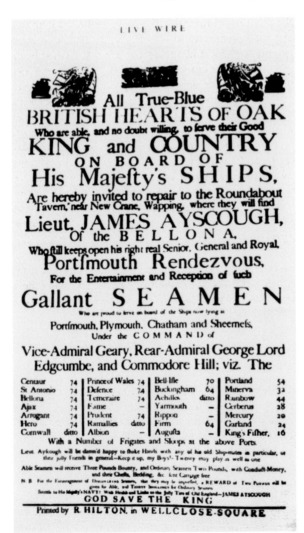

An HMS *Bellona* recruiting notice.
(Collingwood Magazine Archives)

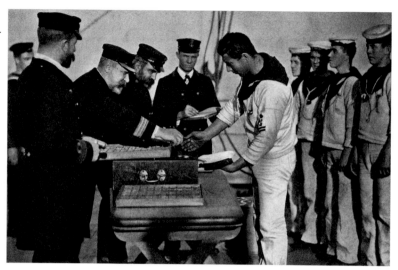

Payday aboard HMS *Royal Sovereign* c.1895. (Naval Museum Portsmouth)

Many ratings wanted the opportunity to gain a naval commission which, since the introduction of dedicated warships, had been awarded by the Admiralty to sea officers of lieutenant rank and above. The significance of this commission was that it identified that sea officers had been given Military Branch status and were qualified to take command of a ship. While a very few selected seamen who were employed in the fighting of the ship had the opportunity as the non-commissioned part of the military branch to gain a commission, other specialist professions were denied it and this was to be a continual source of debate well into the twentieth century. For the emerging engineers who did achieve commissioned status towards the end of the nineteenth century, another source of irritation was the fact that they, along with other non-seaman specialist professionals, were specifically categorised as Civil Branch officers and lacked parity in conditions of service and privileges with the officers of the Military Branch. Interestingly, issues of this nature had been first addressed by the Admiralty in 1731 when improved conditions of service were used to encourage the recruitment of the first gunnery specialists, but few further advances had been made before the advent of steam engineering. As Victorian society evolved in an increasingly technical era, the issues of status and career prospects started to become matters of concern for the rest of the men of the Fleet.

Thus, in addition to funding and implementing technology changes, the Admiralty was now finding itself obliged to consider the service conditions and branch structures that would best support the rapidly changing nature of the Royal Navy's men of war and secure the manpower needed for its ships.

The Impact on Manning

The speed of technical advance during the industrial revolution meant that the Fleet quickly became comprised of ships fitted with a mix of old and new technology. Virtually each hull demanded a unique set of skills from the ship's company and a tailored departmental structure for the effective deployment of those skills. Perhaps it was because of this rate of change and the broad variance in ship construction and outfit that the Admiralty adopted a 'wait and see' approach to implementing such things as shore-based training and the reorganisation of ship board responsibilities. It was not until the late nineteenth century that these matters were felt important enough to be addressed and some were not resolved until after the Second World War.

The task of organising the ship's complement became more and more difficult as the introduction of different technologies, requiring diverse skills, grew rapidly from a trickle to a torrent. While this rate of change started to stabilise after the urgency of the Second World War, it did so at a level that still demanded constant management of manpower and skills in order to keep the Royal Navy at the operational forefront. This is still the situation today.

Managing the Technology

The purpose of this book is to review the way in which electrical technology was introduced into and managed by the Royal Navy up until the establishment of the Electrical Branch in 1946 and then to portray the evolution of the Branch until the end of the twentieth century. It was during this period that the Electrical Branch became more widely known as the 'Greenie Branch' and the officers and ratings forming the Branch became known collectively as 'the Greenies'. The term also became synonymous with HMS *Collingwood*, the shore-based training establishment in Fareham, Hampshire, which became the training Alma Mater for the Electrical Branch.

While the Electrical Branch gave some stability to the provision of electrical engineering support at sea for some 30 years, the unstoppable tide of technology started to force major changes in branch structures in the late 1970s, the late 1980s and the late 2000s. Paradoxically, the first two major changes devolved certain aspects of electrical engineering support back into other branches, thus reversing some of the decisions made leading up to the creation of the Electrical Branch in 1946. However, the most recent change reintroduced an Engineering Branch with unitary electrical responsibility, somewhat ironically echoing the thoughts of the Engineer-in-Chief of the Royal Navy in 1945 during the debate about the creation of the Electrical Branch. It is also interesting to note that all of the cases for further branch restructuring were based on reasons not dissimilar to those existing in the mid 1800s, namely social as well as technological change.

HMS *Collingwood* commissioned as a wartime training establishment in 1940. (Collingwood Photographic Department)

CHAPTER 2

ELECTRICAL TECHNOLOGY 1850–1938

Surface Weapons

From its introduction into naval warfare in the fourteenth century, the cannon was muzzle loaded with the size of projectile constrained by the amount of gunpowder that could be used as a firing charge without the barrel bursting. This limitation was due to the structural weaknesses inherent in barrels manufactured from cast iron. As casting techniques improved and better machining facilities became available, so cast iron barrels were produced which could accommodate bigger charges, with less risk to the barrel, and fire larger cannonballs.

By the early 1800s, the maximum size of cannon being used on board Royal Navy ships was the 32-pounder. Although a 42-pounder had been introduced in the late 1700s, it had been withdrawn from service because its weight meant that too many men were needed to operate the gun and the rates of fire were too low. Although the 32-pounder had a range of around 2,600 yards at an elevation of 8°, according to a range table produced in 1860 by the Gunnery School at HMS *Excellent*, it was normally used at lower angles of elevation and much closer ranges. At the time, cannon was categorised by the weight of cannonball – or shot – fired and the 32-pounder long gun was probably the largest gun that could be fired from a wooden hull owing to the forces being applied to the deck during firings. When the wrought iron barrel was developed in the 1850s, this paved the way for what became the largest cannon of its type, the 68-pounder. In all 26 68-pounder long guns were fitted on the lower deck of HMS *Warrior* in 1861.

As steelmaking technology and precision machining techniques advanced, so gun barrels were produced to much tighter manufacturing tolerances than the earlier cannon barrels. With these advances the concept of the modern naval gun, as opposed to cannon, was introduced, whereby the barrel required the addition of a breech mechanism and mounting arrangement. With the

68-pounder muzzle loading cannon in the HMS *Excellent* drill shed, 1896. (Excellent Museum)

arrival of precision engineered barrels, the traditionally shaped shell replaced the cannonball as a more efficient design for harnessing the energy generated by the charge detonation and firing out to the maximum range possible. This type of gun barrel led to the introduction of a new method of categorising guns by calibre, based on the diameter (calibre) in inches of the bore of the barrel, virtually equivalent to the diameter of the shell being fired from the gun. The weight of both the 68-pounder cannon and the new concept gun were such that the mechanical forces acting on the ship's structure on firing meant that they were only really suitable for mounting in iron-hulled ships.

In the 1770s, the carronade, a much lighter and shorter gun, was invented. It was produced in various sizes which could fire a range of shot from 6–68lbs. The design of the powder chamber was small for the weight of shot, which saved weight and reduced the muzzle velocity causing lower reaction forces on the ship's structure. These features made the carronade suitable for mounting on the upper decks of a wide range of warships, including smaller vessels such as sloops and cutters. The shorter barrel and lower muzzle velocity made the carronade less accurate than the long gun but its heavier shot for size capability made it a devastating weapon at short ranges. Another advantage was that its relatively light weight made it more suitable for manhandling in battle and, when sited on the upper deck, it had a much wider firing arc than the long gun. However, by around 1820, as the ranges of the long gun increased the distance at which naval engagements were being fought, the limited range of the carronade started to make it obsolete as a naval weapon.

With the introduction of the gun port and the increasing weight of cannon, the disadvantages of fighting with muzzle loading weapons was soon recognised and this led to breech loading methods being investigated as early as the sixteenth century. However, the idea did not make much progress as the technology of the times could not achieve a gas-tight seal when closing the breech. It was not until 1859 that Sir William Armstrong developed a breech loading mechanism which, when assembled with his newly designed wrought iron gun barrel, formed the basis for modern naval guns. Armstrong's gun had a barrel with a calibre of 7 inches and it could fire a 110lb shell. Another innovative feature was that spiral grooves, or rifling, had been machined into the internal surface of the barrel. The rifling was designed to exert a spinning force on a lead-coated shell as it passed through the barrel. The spinning induced by the rifling helped to stabilise the shell in the air and improve ballistic accuracy. Eight of the new Armstrong 7-inch breech loading rifled (BLR) guns were fitted in *Warrior*.

Unfortunately, the Armstrong breech proved prone to overheating and it failed to such an extent that a Parliamentary Ordnance Select Committee was tasked to investigate the matter. The Committee reported in 1865 that 'The many-grooved system of rifling with its lead-

68-pounder breech loading cannon in the HMS *Excellent* drill shed, 1896. (Excellent Museum)

An interrupted thread breech block arrangement *c.*1880. (Excellent Museum)

coated projectiles and complicated breech-loading arrangements is far inferior for the general purpose of war to the muzzle-loading system and has the disadvantage of being more expensive in both original cost and ammunition.' Accordingly, the production of the breech loading gun was stopped and the Royal Navy was obliged to revert to muzzle loading guns. While muzzle loading was deemed more reliable, it had the operational impact of reducing the rate of fire achievable from the major calibre guns appearing on the scene. A typical 8-inch shell weighed around 250lbs and this required some form of mechanical ammunition supply and ramming arrangement. Gaining access to the muzzle, either by training over a clear deck or depressing the barrel to a compartment below the weather deck, made the reload process not only more complicated but also more protracted.

In 1880, the interrupted thread breech was invented. This design sealed the breech more effectively than its predecessor and proved to be more reliable in service. As a result, breech loading was re-adopted, but only for guns where rate of fire was deemed operationally critical. This became more applicable to smaller calibre guns which, by virtue of the limited hitting power of a single shell, relied on higher rates of fire to deliver destructive amounts of explosive against surface targets.

As ship hull construction evolved from iron clad to steel and finally to armour plated hulls, so the demand for bigger calibre guns mounted on larger, faster ships increased. From the 7-inch gun of 1859, calibres increased in a series of steps to reach 16.25 inches in the guns fitted in the Royal Navy's Admiral Class battleships built in the 1890s. While this dash for major calibre guns no doubt had some operational justification, there is evidence that it was more often driven by perceptions of the political importance of gun weight and shell size amongst the global sea powers.

HMS *Benbow* with 16.25-inch barbette mounting, 1885. (Naval Museum Portsmouth)

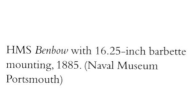

Prior to 1860, naval guns could only fire on fixed – or very limited – bearings with the ship being manoeuvred to bring the guns to bear. These firing bearings were mostly on the beam, but bow and stern chasers were also fitted to give fore and aft channels of fire. What were known as 'transferable' gun mountings with improved arcs of training were gradually coming into use. These guns were usually mounted on manual or steam driven turntables and could be turned to align with gun ports in the ship's hull and, later, in an armoured enclosure for firing to take place. Initially, the enclosures were either 'turrets' or 'barbettes'. The early turret was fixed and fully enclosed the gun with firing arcs being limited by the gun port positions. The barbette was an open topped structure which allowed the gun to be raised for firing with arcs limited only by the ship's superstructure. In due course, the turret itself became trainable and carried the gun giving the benefits of both wide firing arcs and full armour protection. This rotation capability also enabled the correction of convergence at close ranges.[8]

One of the first trainable gun mountings was installed in HMS *Inflexible*. The ship was fitted with four 16-inch guns in twin mountings which were installed midships 'en echelon',[9] to allow at least three of the four guns to be used in fore and aft engagements. The designated main forward firing mounting was on the port side of the ship and the aft firing mounting on the starboard side. Despite a narrow superstructure designed to allow three of the four guns to be brought to bear, it was found that firing ahead or astern still led to considerable blast damage to the ship. The echelon configuration was eventually abandoned in favour of centre line mountings.

Since the introduction of the gun port in the seventeenth century, the broadside engagement at close quarters had been the accepted method of engaging the enemy at sea. The primary reason for this was that while a synchronised firing of all cannon at maximum range was deemed to be the most effective method of engagement, the gun smoke generated invariably caused loss of sight of the enemy and any subsequent firings were only of much use if conducted at point blank range. This range removed any real need to aim the gun and meant that firing was left to the independent control of a local gun captain. Although not technically available at the time, the solution was to have the actual firing of the guns under the direct control of an officer stationed at a remote position which was clear of gun smoke. Such a system would allow broadside engagements at greater ranges, continuous observation of the enemy's position and assessment of the fall of shot for range correction.

HMS *Inflexible* at Malta, 1883. (Naval Museum Portsmouth)

SECTION OF TUBE VENT SEALING ELECTRIC.

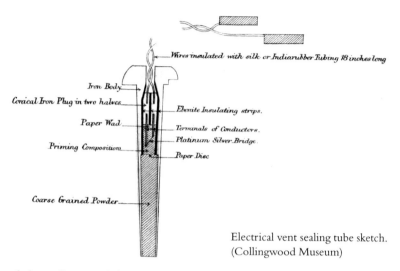

Wires insulated with silk or Indiarubber Tubing 18 inches long

Iron Body

Conical Iron Plug in two halves

Paper Wad

Priming Composition

Ebonite Insulating strips.

Terminals of Conductors.

Platinum Silver Bridge.

Paper Disc

Coarse Grained Powder

Electrical vent sealing tube sketch.
(Collingwood Museum)

It had already been discovered that heat, not just flame, could be used to set off an igniter charge and with the reintroduction of the breech loading gun in the mid nineteenth century came the first major advance in the firing of the gun since the use of the flintlock at the turn of the century. This firing system came in the form of the vent sealing tube (VST) which was inserted into the propellant charge either through the breech mechanism or down through the top of the barrel. The tube contained a method of generating heat which then set off a flash charge that was directed down the tube to initiate the igniter material contained within the main propellant charge. One of the early types of VST was based on the heat produced by the friction generated when a goose quill was pulled through the copper tube by a lanyard. Although mainly used by the British Army until the end of the nineteenth century, the Royal Navy had used this type of tube in muzzle loading cannon but it did not resolve the perennial problem of remote fire control and the synchronised firing of guns as a broadside or, as it was known later, a salvo.

In 1800, Alessandro Volta invented the Voltaic Pile Battery and from this grew the science behind electrical and chemical phenomena known as electrochemistry. The electromotive force, or voltage, of the early batteries was of limited use outside the laboratory. However, by the mid 1800s the voltages being achieved were found to be sufficient to produce a heating effect which had the potential for use in the detonation of explosive material and the development of an electrically operated VST. At last the Royal Navy saw a potential solution to their problem of a remotely operated, electrically powered firing system for use on board ship. Eventually a Voltaic Pile Battery consisting of 120 cells was found to have enough voltage to make it suitable for the task of quickly heating a fusible link and initiating the igniter material inside a charge of gunpowder being used as a propellant.

It was in 1868 that Commander J.A. Fisher – who already had a reputation as a technical innovator following the publication of his book on electrically detonated torpedoes – produced an appendix to his book which proposed a design for an electrical gun firing system. ('Jackie' Fisher would have a major influence on warfare technology development for the next fifty years. He became First Sea Lord in 1904.) His idea led to a standard specification for the installation of electrical firing circuits to enable synchronised firing. Two electrical cables were run along the deck head over the gun positions with one connected directly to one pole of the firing battery and the other connected through a firing switch to the opposite pole. Branch cables were connected to each deck head cable and were brought down to each gun for connection to an insulated block. When the gun was loaded, the VST was inserted and connected to the block ready to receive the firing signal. Thus remotely controlled broadside firing came into being.

Perhaps the earliest formal recognition of the naval application of electricity appeared when the Admiralty issued a Confidential Circular, numbered 13S and dated 27 February 1873:

> In future the electric firing apparatus [for guns] supplied to Her Majesty's vessels is to be fitted as soon as the ship is placed in the 1st Division of the steam reserve, under the immediate superintendence of the Commander or Lieutenant of the Portsmouth Naval Torpedo School (timely information being afforded to the Captain of the *Excellent*) and the fittings so arranged are not to be subsequently altered without the sanction of Their Lordships.

This circular went on to give some detail as to how the firing circuits should be run throughout the ship, including the position, insulation requirements and testing of the firing key. Although the responsibility for installing the firing circuits lay with the Torpedo School, the responsibility for carrying out pre-firing checks on board lay with the Gunnery Officer. This duty would have been one of the first documented user checks and the relevant Admiralty Order read as follows:

> The electrical firing wires are frequently to be tested by the Gunnery Officer and invariably before use, either with the galvanometer and test battery, or, if the firing battery is built, by keeping the firing key pressed down and sending a trustworthy man to each gun, with directions to make the necessary connections and take the shock.

It is interesting to note that the check required the assistance of a 'trustworthy' man, notwithstanding the fact that such a man would have previously found out that he could expect an electric shock of around 80 volts if the firing circuits tested correct!

As the weight of guns and their mountings increased rapidly, mechanical design and manufacturing limitations meant that the machinery was prone to imbalance and needed considerable force to impart training and elevation movement. This meant that the use of steam soon became the only real option for moving heavy guns, particularly with the elevating and running out of the guns to their firing position. In 1874, the forward mounting in HMS *Thunderer* was adapted for hydraulic operation using water as the medium. The performance of the gun was compared with that of her aft mounting; which was manually operated. The results of the evaluation trials came out in favour of hydraulically powered operation, not least because of the fact that 28 men were required to work the hydraulic gun as against 48 men for the manual version.

As improved gun designs requiring less motive force began to appear, so electrically powered mountings started to be developed by, amongst others, the United States. Little interest was shown by the Admiralty but *Vernon* Torpedo School did undertake a study of electrically powered mountings in use with other navies. In 1894, this led to a trial being carried out using the forward 10-inch gun mounting in HMS *Barfleur* with the right gun elevation motion being modified for electrical drive. The results showed that the gun would badly overshoot the required elevation, despite electrical cut outs being operated.[10] *Vernon* also carried out trials on a smaller 6-inch gun fitted in HMS *Seagull,* a torpedo gunboat. Both training and elevating motions were modified and set up to be controlled using a single switch. Speeds of 7° per second in training and 4° per second in elevation were achieved. Overshoot of the fire position was prevented by shorting out the armature of the shunt-wound, 80-volt DC motors. These trials were more successful and the speed of motion sufficiently demonstrated that electric drive trains were suitable for smaller gun calibres where the emphasis was on rate of fire. While gun manufacturers still dabbled with electrical power drives in up to 12-inch gun mountings, hydraulic drives remained the preferred technology for major calibre gun operation in the Royal Navy until after the Second World War when the air threat started to demand higher rates of fire to make up for low hit probability rates and gun calibres started to reduce with the demise of the battleship.

Towards the end of the nineteenth century, power for electrical firing circuits started to be provided from low power switchboards with battery power only being used in an emergency. Throughout this period, the control features of the firing circuit continued to evolve as the needs of the gun mounting also changed. Notable features included safety interlocks – to avoid the dangers of self inflicted damage when training gun mountings – the introduction of centre line turrets and, in 1915, gyro interlocks to ensure that the firing circuit was only closed when the ship's roll and pitch attitude coincided with the horizontal plane being used to calculate the firing solution.

From around 1890 and before the advent of aircraft and submarines, the major close engagement threat to Royal Navy's Battle Fleet was perceived to be from torpedo boats. This resulted in smaller calibre guns, principally 4-inch and 4.7-inch, being fitted as secondary armament in capital ships. These guns used fixed ammunition, a combination of charge and shell into one round,[11] that enabled quicker reload times and manhandling during the loading process. Priority was given to achieving a high rate of fire with the smaller explosive payload being deemed sufficient to inflict lethal damage on unarmoured vessels. However, the gun mountings were designed with only a limited elevation capability as they were intended to meet the surface threat at relatively close range. This later proved a limitation in their use against any airborne threat. As major Fleet units moved further away from the coastal waters, so the torpedo threat to the Fleet transferred to the torpedo boat destroyer (later destroyer) which could operate in the open ocean, carried torpedoes and was also fitted with 4-inch guns as the main armament.

At this point, it is worth recounting some of the history of the political environment for technology development in the early twentieth century. It was this environment that not only affected the pattern of technology investment but also the development of the personnel skills required to support that technology.

In October 1904, 'Jackie' Fisher was appointed as First Sea Lord and set about implementing major reforms of the operational, materiel and training organisation of the Royal Navy. A significant part of these reforms was to streamline the Fleet, which Fisher considered had become operationally unwieldy and too expensive to maintain owing to the reluctance to take obsolete ships out of service. The main reason for this reluctance was the long established policy of maintaining a fleet which, numerically, was the equivalent of any two other world navies combined together. This arbitrary measure, known as the Two Power Standard, had been re-established in law under the 1889 Naval Defence Act and led to unsustainable rises in the annual Naval Estimates to support what Fisher regarded as a largely 'useless' fighting capability.[12] This assessment was widely debated and had many detractors both in and outside the Royal Navy but Fisher remained determined to see his ideas implemented.

One result of Fisher's technical reforms was to consolidate capital ship main armament about the 12-inch gun mounting and this concept, coupled with an operating speed of 21 knots, formed the core fighting capability requirement of HMS *Dreadnought*. *Dreadnought* commenced build in November 1905 and the name became the iconic generic term for similar classes of battleship developed by other navies which subsequently followed this change in Royal Navy operational policy. In conjunction with a build programme for more Dreadnoughts and modernised armoured cruisers, Fisher also rationalised the global disposition of Royal Navy Fleets with the focus on a Grand Fleet based in Britain and ready for the war with Germany. Details of Fisher's significant reforms in the personnel area are covered elsewhere in this book.

HMS *Dreadnought* in 1906.
(Collingwood Museum)

After six years as First Sea Lord and setting the strategy for the Royal Navy in the face of the German threat, Fisher retired in January 1911 at the age of 70. Not all naval officers had shared his views, particularly those outside the 'Fishpond',[13] and some were glad to see him leave having felt that his tenure had been autocratic and, in certain areas, blinkered. However, during Fisher's time, the German warship building programme had become so intensive that his strong beliefs about containing the threat had taken Britain into a costly and escalating naval arms race in order to maintain its superiority. Such was Fisher's concern about the need to modernise the Fleet, it is probable that he arranged for Admiral of the Fleet Sir Arthur Wilson to succeed him as First Sea Lord, partly because Wilson shared Fisher's strategic view and partly, it is said, because he had enough seniority to keep the serving Admirals who still disagreed with Fisher under control.

The rising expenditure was not well received by the Liberal government of the time, not least by Winston Churchill who was then the Home Secretary. In October 1911, Churchill was appointed as First Lord of the Admiralty, probably being expected to cut back on Fisher's expansionist aspirations. Instead, Churchill struck up a relationship with the retired Fisher and became more convinced by the strategic thinking of the aging Admiral. Notwithstanding this relationship, Churchill's arrival at the Admiralty quickly saw the departure of Wilson as First Sea Lord in November 1911. Wilson was replaced by Admiral Sir Francis Bridgeman, who resigned in December 1912, allegedly under pressure from Churchill, and was replaced by Admiral His Serene Highness Louis of Battenberg. Thus in the four crucial years before the outbreak of war, the Royal Navy had seen four different Admirals as First Sea Lord in charge of its affairs.

Following the outbreak of the First World War on 3 August 1914, the Royal Navy's early wartime operations were not wholly successful. The global, but aging and thinly spread, British Fleets failed to stop German commerce raiders from deploying to attack British overseas trade routes and three cruisers, HM Ships *Aboukir*, *Cressy* and *Hogue* were torpedoed by a single U-boat and sunk off the coast of Holland in September 1914 with the loss of over 1,400 lives. Churchill brought back Fisher as First Sea Lord to replace Louis of Battenburg in October 1914.[14] In November 1914, two cruisers, HM Ships *Good Hope* and *Monmouth,* were outgunned and sunk at the Battle of Coronel and, in February 1915, HM Ships *Irresistible, Inflexible* and *Ocean* were lost to mines during the ill-fated Dardanelles Campaign. The Germans also declared unrestricted submarine warfare in February 1915 and British shipping losses were rising rapidly in the face of increasing U-boat operations. The only slight relief for a dismayed public opinion was the victory off the Falklands Islands over the German victors at Coronel.

In his typically robust fashion, Fisher argued his concerns with an equally vigorous Churchill regarding the naval strategy for the war, particularly the naval role in the Dardanelles Campaign.

HMS *Furious* after her forward deck conversion to carry aircraft in 1917. (Imperial War Museum)

These differences were often made public and eventually led to Fisher's resignation in May 1915. Churchill was forced to resign later that month when the coalition Liberal-Conservative government was formed and he was replaced by the Conservative, Arthur Balfour. Admiral Beatty, at one time Naval Secretary to Churchill, was alleged to have said that 'The Navy breathes much freer now it is rid of the succubus Churchill,' a pointer to the personality and strategic differences then rife at senior levels in the Royal Navy. This situation inevitably added to the earlier lack of strategic coherence and technology direction at the centre, with resulting war outcomes being further prejudiced by weak communication with the Admirals in charge of the operational Fleets. Fisher was replaced by Admiral Sir Henry Jackson, another technical innovator in the Fisher mould, whose enormous contribution to radio communication technology in the Royal Navy is covered later.

Naval technology weakness had become quickly evident during the First World War, particularly in the areas of ship armour, magazine construction and anti-submarine warfare. Perhaps in response, Balfour set up the Admiralty Board of Invention and Research (BIR) in July 1915.[15] The BIR was set up to be independent of the Admiralty but was granted access to all technical data held by any of the Admiralty departments. The BIR was composed of six research sections, manned by civilian scientists and engineers. Fisher was called out of retirement to chair its management board.

The creation of the BIR was no doubt laudable as it provided a vehicle to get many noted scientists involved in supporting the war effort, something which had been eagerly sought by the scientific community as a whole. However, its true function was never clearly defined and through its title, namely Invention and Research, it became all things to all men ranging from pure research, through invention to the task of screening ideas to help win the war. From the uniformed Admiralty point of view, there was the negative perception that the BIR's activities were under the control of Fisher and this did not bode well for any working relationship with the Royal Navy's wartime operations staff.

Following the pattern of the *Dreadnought* design conceived under Fisher's first term in office, the 12-inch gun remained the battleship main armament of choice until the introduction of the *Queen Elizabeth* Class. This class started build in 1912, was fitted with eight 15-inch and twelve 6-inch guns as well as submerged torpedo tubes. Termed 'Super Dreadnoughts', the ships had steam turbine propulsion systems and could make up to 25 knots. Gun calibre finally reached a British peak when Fisher, in 1915, backed the development of the 18-inch Mk1 gun. This turned out to be the Royal Navy's biggest gun but only three were ever manufactured. Originally, two of the guns were to be fitted in HMS *Furious* but the forward one was replaced during build by a platform to be used for the launch of aircraft. During acceptance trials, the recoil of the aft mounting was shown to cause too much stress on the hull and it was removed. This allowed the flight deck to be extended towards the stern of the ship and safer operation of aircraft.

After the First World War, the 1922 Washington Arms Limitation Treaty globally restricted the size of gun calibre and no further 18-inch installations were permitted under international law. This could have been considered academic, as by now gun ranges were exceeding the limits of surface target detection and tracking. Also, the growing air threat from torpedo and bomb carrying aircraft was not effectively being addressed by the big gun. In particular, the limited elevation capability, low rates of fire and pre-set methods of fuzing shells meant that tons of high explosive could be fired with little or no hope of hitting an aerial target. The answer at the time lay in the smaller calibre, shorter range, higher rate of fire gun.

The British Army adopted machine guns for the land battle in the late nineteenth century. Initially, these guns were of the hand cranked variety but, in 1900, the British Army deployed the recoil operated and more reliable 37mm Vickers-Maxim gun during the Boer War. As the potential air threat emerged in the buildup to the First World War, the Army converted the Vickers-Maxim for anti-aircraft defence and it was used both in the trenches and for Homeland defence. The Vickers-Maxim was a British-built version of the Nordenfelt-Maxim which had been licensed for building throughout the world prior to Vickers buying out the Nordenfelt half of the company. On the other hand, the Royal Navy did not appear to consider that aircraft would seriously trouble its Fleet until February 1914, when the air-launched torpedo attack became a reality. Thus it was not until 1915 that a 40mm version of the Army's Vickers Maxim started to be fitted to cruisers and destroyers. The guns were fitted to high elevation angle mountings and had a rate of fire of 200 rounds per minute with an effective range of 1,200 yards. The 40mm (2lb) shell was deemed adequate for defence against aircraft, which were then relatively slow and fragile. The gun was soon dubbed the 'pom-pom' due the sound of the firing action. The 0.5-inch Vickers machine gun was also fitted to ships during the First World War. This weapon fired up to 600 rounds per minute but the much smaller shell meant that it had limited stopping power.

The appearance of the aircraft carrier, and its enhancement of the air threat to the Grand Fleet, prompted the Royal Navy to develop a multi-barrelled version of the 40mm pom-pom. A sixteen-barrel version was produced for installation in cruisers and above by 1930. Single- and four-barrel versions for smaller ships were introduced by 1935. The intention was for the multiple barrel guns to set up barrage fire at a fixed range in order to distract the pilot in his attack approach. Apart from the excessive weight of the larger mountings, another weakness of the pom-pom proved to be the lack of a tracer round. This reduced the aimer's ability to correct the line of fire and meant that the pilot was blissfully unaware of the approaching shell and therefore undeterred.

Gunnery Fire Control

While the problem of synchronised firing had been resolved by the arrival of the voltaic pile battery and the VST, experience from the bombardment of Alexandria in 1882 showed that the passing of gun control information by word of mouth from a remote control position above decks was not practical in the noise of battle, as reported after the Alexandria action by Captain Henry Fairfax, Commanding Officer HMS *Monarch*:

> In the din of the action it was most difficult to be heard through the voice pipes and it was found necessary to pass word to the turrets by officers specially stationed for that purpose. The want, therefore, of some simple, quick and efficient means of communicating with the interior of the ship was, from the conning tower, much felt.[16]

Various designs for the electrical transmission of gun training and elevation orders were tried out over the subsequent years but it was not until 1900 that Barr and Stroud produced an electro-mechanical design which was deemed suitable for widespread fitting in the Fleet.

While target range, bearing and elevation information, as seen from a remote position, could now be made available at the gun mounting, the problem then became pointing and firing the gun at the elevation needed to achieve the trajectory, and hence range, required to hit a target which had yet to be seen by the gun layer and trainer. In local control, aiming had been managed by having training and elevation sighting telescopes which were collimated with the bore of the barrel and separately operated to follow the target, manually trying to compensate for the ship's movement. When in remote control, the operators used the received information to search for the target and point the sights directly at it. It was then the gun layer's task to super elevate the gun to achieve the required range sent from the remote source. Other than in flat calm conditions, ship motion would mean that the gun could only be trained and laid at an elevation to hit the target twice in each roll cycle of the ship. Thus, as the target became simultaneously sighted in the crosshairs of both the training and elevation sights, with the barrel being offset to compensate for the super elevation required to achieve the correct trajectory, so the gun could be fired. Fall of shot spotting could then be used to indicate changes needed to the super elevation angle before the next firing opportunity.

The development of an ability to transmit positional information led to an initial proposal by Captain Percy Scott for director firing of a battleship's main armament in 1905.[17] This involved setting up a master gun sight, or director, as high as possible on the ship and using it to provide continuous training and elevating information to the guns and to fire them at a preset elevation.

A sketch of Percy Scott's director firing system. (Excellent Museum)

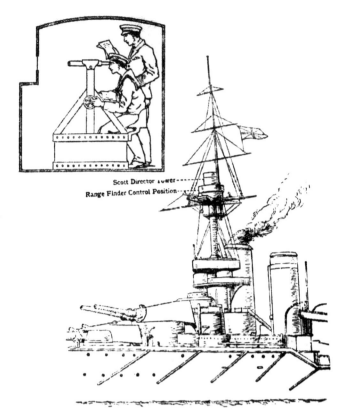

Scott Director Tower
Range Finder Control Position

Such a system allowed all guns in the main armament to simultaneously fire at the target during the roll cycle but at an arbitrary range. However, as all shells in the salvo would arrive on target at the same time, this made observation of the fall of shot from the director easier and for corrections to be applied. This system removed the need for local sighting of the guns, except in an emergency, and allowed command and control from a position much less affected by sea spray, gun smoke and the noise of the turret. Another major operational advantage was that the director sighting could be linked to searchlight operation and this offered a night firing capability not available with local gun control.

The targeting advantages of director firing in this basic form were not widely accepted and, although Scott had implemented some ad hoc arrangements in ships under his command, the system was not formally trialled until May 1910 in HMS *Bellerophon*. The trial proved inconclusive and the Commanding Officer, Captain Hugh Evan-Thomas, did not recommend its adoption by the Fleet. The Director of Naval Ordnance concurred with the recommendation but appointed Scott to work with Vickers to examine any improvements to a concept which had its appeal but lacked the facility of providing continuous gun aiming at the target.

In 1906, the idea of the transmitting station (TS) – a position below decks which acted as the link between the upperdeck control position observing the target and the gun mountings – was introduced. It was in the quiet of the TS that an increasing number of machines were, in due course, employed to generate target position and rates of movement and, after applying ballistic corrections, to compute the training and elevation orders for each gun such that the probability of the shell hitting the target was maximised.

By 1913, thanks to Scott's efforts, along with Vickers who had supported the original trials, a more sophisticated director firing solution which offered continuous aiming and synchronised firing was developed. The system linked the director to the gun via the TS, where aiming corrections could be applied to take account of the relative positions of the director and each of the guns in ship. Despite this advance, Scott still had difficulty in getting any urgency behind a Fleet installation programme, something which he blamed on the bureaucratic approach of the Admiralty and a reluctance to accept ideas that cast aspersions on existing Royal Navy capabilities. In his memoirs, Scott notes that at the outbreak of the First World War only eight battleships were fitted with director finding despite the Commander-in-Chief of the Atlantic Fleet, Admiral John Jellicoe,[18] having witnessed the superiority of the system during a competitive trial between HMS *Thunderer* and HMS *Orion* in November 1911, the former being fitted with director finding. The principles of director firing were eventually widely adopted and evolved rapidly for the control of smaller calibre guns against shorter range surface and air targets. Further major advances in the capability of director control systems came about with introduction of gyro-stabilised gun mountings around 1915, which allowed the rate of fire to be independent of the ship's movement, and the emergence of fire control radars during the Second World War which could lock onto a target and track it to give a continuous fire control solution.

One of the most important machines installed in the TS was the Dreyer Fire Control Table (DFCT), an analogue fire control computer first developed in the years before the First World War by a gunnery officer, Frederic Dreyer, for use in surface firings.[19] The DFCT Mark 1 was entirely manual with target range and bearing information from observers being plotted against time to generate target rate information. The rate information was used in conjunction with the ship's course and speed to generate a gun point of aim which reflected the target's predicted position allowing for the time of flight of the shell. A number of DFCT models were developed over the years and they gradually incorporated more automation due to the advances made in gyro and ship's data transmission technology in the 1920s. When the 'M' Type and, ten years later, the 'Magslip' electrical transmission systems were invented, this led to more accurate data being made available and fully enabled the basic architecture for all modern gunfire control systems in surface mode.

Dreyer's analogue solution to the fire control problem proved robust and effective, if somewhat complex, and two much improved versions of the DFCT saw service in HMS *Rodney* and HMS *Nelson* during the Second World War. However, during the 1930s a less complicated version, designated the Admiralty Fire Control Table (AFCT), began to replace the DFCT in most capital ships. A more limited version of the AFCT was fitted in destroyers and below and these computers were designated as either Admiralty Fire Control Clocks (AFCC) or Boxes (AFCB). As with the DFCT, the AFCT generation of British fire control computers plotted visually acquired target information against time as a means of establishing the range and bearing rates but, in addition, corrections for ship's motion, ship's head and other ballistic factors were now automatically applied to predicted target bearing and range for the calculation of the fire control solution.[20]

While the Dreyer Fire Control Table and its successors appeared adequate for dealing with the existing surface threat, they did not resolve the growing problems of the submarine and aircraft threats which, from the early twentieth century, started to emerge with a torpedo launch capability being built into both platforms. In particular, the first launch of a torpedo from an aircraft by Captain Alessandro Guidoni of the Italian Navy in February 1914 heralded the arrival of a threat which, even at the low air speeds of the time, was quickly to prove beyond the capability of the optically based analogue methods used in the DFCT, AFCT and AFCB. As aircraft performance continued to improve after the First World War, ship anti-aircraft defence became of much more concern but, within the context of severe post-war financial constraint, there was little progress towards finding a solution.

The emerging threat – a Short 184 seaplane carrying a 14-inch torpedo. (Collingwood Museum)

In 1926, the Imperial Defence Committee produced a report which postulated that, in order to carry out precision torpedo or bombing attacks on a ship, an aircraft would have to fly straight, level and at a constant speed during the weapon delivery approach phase. With these restrictive parameters in mind, the Admiralty contracted Vickers Ltd to develop a high angle control system (HACS) for the control of guns in the air defence of capital ships, aircraft carriers and cruisers. Smaller ships were to be provided with a more limited version which was eventually designated as a Fuze Keeping Clock (FKC).

The Vickers fire control system design model was a child of its time. Its functional basis was the engagement sequence as initiated by the Gunnery Control Officer in the HACS director and an attacking aircraft flight path being straight, level and at constant speed. The Control Officer was tasked with assessing the aircraft heading using a binocular graticule and, based on the type of aircraft, estimating its probable speed. Three rating operators were also stationed in the director; one was charged with measuring target bearing, a second measured target elevation and the third took target range measurements using an optical coincidence rangefinder. Before the advent of gyros, all three operators had to compensate for ship motion and this made the successful operation of the rangefinder almost impossible in any sea.

Information from the above decks director was passed to HACS Table, situated below decks in the TS, where the data was mechanically input by yet more operators into an aircraft equivalent of the DFCT. These inputs were then processed by table mechanisms designed to carry out analogue computation of the gun training and elevation angles required to fire a shell at the aircraft's future position and the time to be set on the fuze for that shell to explode when in coincidence with the aircraft. Corrections would be applied at the table, based on any target motion data and spotting of shell bursts from not only the director but also the bridge. Unfortunately the initial assumptions were so liable to human error that, when combined with operating error, the kill probability of such a system was very low.

The HACS, as described, was the basis for the Royal Navy's air defence going into the Second World War. The weaknesses in fire control assumptions, director operation and range finding were recognised pre-war but despite the emergence of alternative optical systems, more accurate power drives and data trains, gyro rate aiding and stabilisation technology, none of these improvements had reached the Fleet by 1939. The anti-aircraft performance of HACS was eventually to be seriously tested in 1940 during the early naval engagements off the coast of Norway. The known shortcomings were severely exposed by the German use of dive bomber tactics, which presented attack parameters that were totally different to the theoretical parameters used as the basis for the HACS fire control solution.

Underwater Weapons

The first documented underwater weapon was a floating mine. It was invented by David Bushnell and used by the Americans during the 1776 Revolutionary War. Bushnell's mine comprised a watertight wooden keg, loaded with gunpowder and suspended from a float which allowed it to drift with the current in the direction of enemy shipping. Detonation was by means of a flintlock pistol arrangement which relied on physical contact with a ship's hull. The keg mines were crude in operation and proved too unpredictable to be very successful. However, the concept of the sea mine was borne and further development of the principles across the world soon led to the use of statically moored versions which were mechanically detonated and used to deter ship movement.

In addition to his mine, in 1775 Bushnell also built a one-man submersible, christened the 'Turtle' and arguably the first submarine, for approaching enemy ships underwater and attaching an explosive charge to the hull. Although Bushnell also referred to this charge as a 'torpedo', it was more akin to an early limpet mine in that it was mechanically, rather than magnetically, attached to the wooden hull using an augur device to drill into the hull in order to secure the explosive charge. There was one documented attack against a British warship, HMS *Eagle*, in September 1776. According to the Americans, the attack was only unsuccessful due to the augur failing to

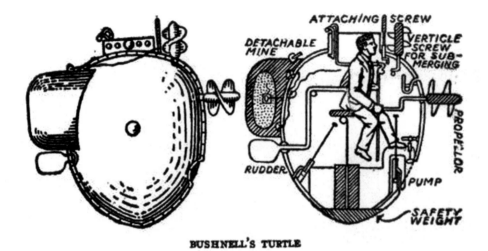

BUSHNELL'S TURTLE

Bushnell's Turtle submarine *c*.1776. (Collingwood Magazine Archives)

penetrate the ship's copper bottom, by then a feature of many British men-of-war. However, unfortunately for Bushnell, records show that the *Eagle* was not copper sheathed until around 1783. The failure was probably due to the inability of the single man crew to pedal the propulsion screw, manually operate the depth control and ballast system and attach the torpedo all at the same time.

The Admiralty initially showed little interest in developing mines for the Royal Navy because they were viewed as defensive weapons, with static mines being primarily associated with harbour protection. There was no thought of such weapons being used for offensive operations such as the blockading of an enemy harbour. This attitude was rooted in the British tactics employed around the turn of the eighteenth century by, amongst others, Lord Nelson. At that time, the Royal Navy was powerful enough not to feel the need to be protected in its own harbours and preferred to use ships to patrol outside enemy harbours inviting opposing naval forces to put to sea and do battle. Alongside this policy, there was an entrenched view that mines were 'an objectionable mode of warfare' and 'repugnant to men of our own race'.[21]

Some of the first static mines used mechanically activated chemical detonators consisting of lead horns containing a glass ampoule of sulphuric acid held within a mixture of potassium chlorate and fine sugar. This type of mine still relied on ship contact, at which point the horn was crushed and its contents underwent a chemical reaction which produced sufficient heat and flame to set off a main charge made up of gunpowder packed in wooden casks.

In addition to mechanically detonated mines, the advent of the battery and the dynamo brought about mines which were designed to be electrically detonated via cables from the shore. There were two main classes of electrical mines developed by the mid 1800s: the 'observation' mine and the 'electro-contact' mine. The observation mine was laid under the surveillance of shore defensive forces and connected to a shore-based electrical power supply, either battery or dynamo. When enemy shipping was sighted in the vicinity, any adjacent mine could then be detonated from the shore, without any ship contact.

The electro-contact mine was suited for minefields remote from an observer, but ship contact was necessary. The mine would be wired to the shore and when mechanical contact was made the collision would operate a mercury switch which would complete an electrical circuit within the mine and sound an alarm bell ashore. On hearing the alarm, the firing circuit for the mine would be completed from ashore and detonation would take place. The advantage of these two types of mine was that detonation could be prevented when friendly shipping was in the area of the minefield.

Controlled mines being made ready, Wei-Hai-Wei, 1905. (Naval Museum Portsmouth)

Controlled mine detonation. (Naval Museum Portsmouth)

During the Crimean War (1854–1856), the Russians laid down static minefields using a mix of mechanically and electrically detonated mines to defend the Baltic harbours of Sveaborg and Kronstadt and their Black Sea ports. The presence of these mining defences did much to thwart British attempts to take Kronstadt, the fortress guarding the seaward approach to St Petersburg. British intelligence knew of the existence of mines and in June 1854 one was recovered and exploded under trial conditions to reveal the weapon's true potential as a threat. The combination of mines and shallow waters, overlooked by the long-range shore batteries, eventually dissuaded the British commander, Vice Admiral Sir Charles Napier, from committing his capital ships to a naval bombardment and he withdrew his fleet. Napier was later relieved of his command for failing to press on with his mission. In 1855 under Admiral Dundas, the British returned to Kronstadt and suffered damage to at least four ships as a result of controlled mine explosions. These events, arguably, led to the first recorded British attempts to sweep enemy mines on a large scale and more than 50 were recovered. While the damage did not result in the loss of any ships, probably due to the small size of the explosive charges, the psychological impact was such that the Admiralty were finally convinced that this form of 'ungentlemanly' warfare could not be ignored. By 1860, underwater weapons had become an important means of defence used by many countries around

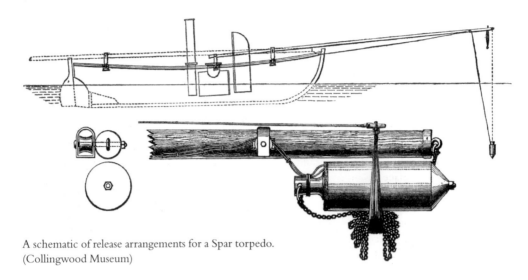

A schematic of release arrangements for a Spar torpedo.
(Collingwood Museum)

the world and the Royal Navy started to take mine warfare, particularly countermining, more seriously.

The early 1860s heralded the arrival of the first electrically detonated weapons to be installed in ships, the spar – or outrigger – torpedoes. These torpedoes were not in the form of the modern torpedo but more like a transportable mine comprising a guncotton charge placed at the end of a long spar and deployed over the bows of a steam pinnace. The tactical concept was to place the pinnace close enough to an enemy vessel to enable the charge to be dipped under the water and for contact to be made below any armoured belt. The warhead was then electrically detonated by the officer in command using a voltaic pile type battery consisting of sheets of alternate copper and zinc separated by layers of 'fearnought'[22] material dipped in a mixture of vinegar, salt and water. Fisher's 1868 book *Electricity and the Management of Electric Torpedoes* recalls that the Royal Navy's Gunnery School, HMS *Excellent*, had both a steam launch and a gig fitted out for outrigger torpedo trials. Fisher records that the gig's torpedo arrangements were constructed 'to admit of 50lbs of gunpowder, submerged by nine feet and at seventeen and a half feet from the stem'. He also states that trials showed that it could be 'fired with perfect safety to the boat and its crew'.[23] Although probably equally hazardous to both the user and the enemy,[24] the outrigger torpedo system raised the awareness of a class of weapon that was to significantly affect the development of future naval electrical applications.

Spar torpedo test firing in Portsmouth Harbour, 1868. (Naval Historical Branch)

Spar torpedo boat in Portsmouth Harbour, 1868. (Naval Historical Branch)

The importance of the sea mine as a weapon was further increased when, in 1866, Dr Otto Hertz developed an autonomous electro-mechanical detonator, known as a Hertz Horn. The horn comprised a lead tube inside which was a glass tube filled with a potassium bichromate electrolyte solution. The glass tube was suspended over carbon and zinc plates at the base of the horn. When the lead horn came into contact with a solid object, such as the hull of a ship, it bent the lead tube causing the glass tube to fracture. The solution was released onto the plates, thus forming a battery capable of producing 1.5 volts and that was sufficient to fuse a fine platinum wire embedded in the mercury fulminate of the detonator. The detonator then initiated a guncotton primer so it fired the main charge. The advantage of the Hertz Horn was that it was remarkably reliable and could survive many years in the ocean, thus leading to the adage that once in place or, more hazardously, when drifting, the contact mine was a weapon that never gave up until it was destroyed.

Even with these advances in mine technology, as far as the Royal Navy was concerned the nature of mine warfare was still defensive and it had limited potential for use as an offensive weapon. This thinking led to more concentration on the development of countermining capabilities in order to gain freedom of movement for the British Fleet. The Torpedo School was finally instructed to withdraw electro-mechanical mines from service in 1894 and mechanical mine development came to a virtual halt with no work being done to progress any automatic depth setting capability, an essential feature for offensive mining.

This mine warfare policy remained in force until 1904 when, during the Russo-Japanese War, both sides demonstrated the effective use of mines in an offensive role.[25] This came about with the development of mines which could be quickly laid and automatically placed at the required depth without any prior survey of the sea bed. Such mines were used to quickly close harbours or shipping lanes as a tactical move to stall or intercept the enemy fleet by laying them within hours of earlier clearance action or free passage. In 1905, with these lessons in mind, a rapid expansion in the number of fast mine-laying ships and mines was ordered by the First Sea Lord, Jackie Fisher, in preparation for a war with Germany.

In 1866, the Englishman Robert Whitehead conceived the idea of using an automotive torpedo to deliver an explosive charge from a remote position to a point below a ship's waterline in the belief that the damage caused would be greater than that achievable by above water bombardment. Whitehead developed his idea into a production model for the Austrian Navy.

In 1869, the Royal Navy visited the Whitehead trials site in the port of Fiume, then in Austria, following a report from the Commander-in-Chief Mediterranean Fleet about the effectiveness of Whitehead torpedoes. In 1870, the Admiralty commissioned Whitehead to demonstrate his

Automotive torpedo test firing in Portsmouth Harbour 1890. (Museum of Naval Firepower)

Automotive torpedo
test firing in 1870.
(Museum of Naval
Firepower)

14-inch and 16-inch torpedoes with launches from HMS *Oberon*, a sloop that had been fitted
with a submerged tube in the bow from which both torpedo types could be discharged using
a ram driven by compressed air. Over 50 runs were carried out using the 14-inch torpedo and
the trial was repeated using the 16-inch version. Respectively, the average speed and range for
each weapon was 8.5 knots over 200 yards and 7.5 knots over 600 yards. The final trial was to
fit the warheads and fire both weapons at the captured French hulk *Aigle* from a range of 136
yards. The 16-inch was fitted with a 67lbs guncotton charge and it blew a hole in the side of
the ship. The 14-inch carried an 18lbs charge and exploded when caught by a net which had
been hung over the side to protect the ship from further damage.[26] The Admiralty was suitably
impressed and ordered a batch of torpedoes for the Royal Navy. This led to the French, German
and Chinese navies following suit and Whitehead started to export the torpedo around the world,
by 1881 achieving sales approaching 1500 weapons. In 1871, the Admiralty bought the rights to
manufacture Whitehead's 16-inch torpedo at the Royal Gun Factory (RGF) at Woolwich under
the design control of the Superintendent of the Royal Laboratory. A British-built variant of the
16-inch weapon came into service in 1872 and introduced a number of important improvements.
These included twin contra-rotating propellers for straighter running and a higher pressure for
the compressed air propulsion system to increase the weapon speed to 12 knots. Whitehead duly
incorporated the British design improvements into his 14-inch model and increased the warhead
to 26.5lbs. In 1877, the production of the 16-inch at the RGF was stopped, presumably due to
the increase in size of the warhead in the smaller variant. In 1890, torpedo design authority was
transferred from the Royal Laboratory to the RGF and, based on Whitehead's earlier development
lead, an 18-inch weapon went into production.

By 1881, the Royal Navy had over 40 torpedo gunboats in service with some capability for
either automotive or outrigger torpedo operations. British shipbuilders were also exporting
these gunboats throughout the world as the adoption of the torpedo spread. It was a British-
built gunboat, sold to Chile, which is believed to have achieved the first sinking of a ship with an
automotive torpedo in April 1881. The weapon was a 14-inch Whitehead variant and it was fired
at 100 yards to sink the Chilean ironclad, *Blanco Encalada*, during the Revolutionary War.

The emergence of the torpedo also caught the eye of the British Army as the Woolwich Gun
Factory was under their control. In 1886, this led to the Royal Engineers being put in charge
of operating shore based torpedo systems for harbour defence. These systems were installed at
a number of naval harbours, both at home and abroad. The weapon adopted was the Brennan
Torpedo,[27] a wire-guided weapon with a range of some 3000 yards at a speed of around 20 knots.
By 1905, the Brennan's limited range and the problems of using it at night caused the Army

Preparing to launch a Whitehead torpedo from HMS *Thunderer*, 1872. (Naval Museum Portsmouth)

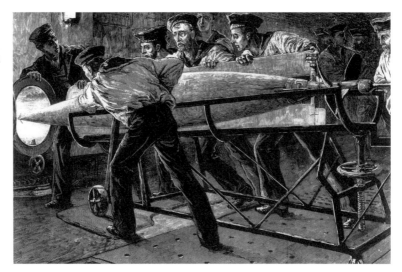

to recommend that its use as a component of fixed harbour defences should be discontinued. This coincided with the transfer of responsibility for seaward harbour defence from the Royal Engineers to the Royal Navy.

The Whitehead torpedo used a compressed air driven propulsion system but alternative systems were also developed with a view to increasing the range and speed of the weapon. Electrical propulsion was tried around 1873 by Ericsson, using power sent down trailing cables, and later by Thorsten Nordenfelt, who developed a battery-driven version but retained the trailing cables as a means of sending electrical impulses to the weapon for guidance purposes. Steam and rocket propelled torpedoes were experimentally produced in the United States but the Whitehead compressed air system provided the development basis for Royal Navy ship-borne applications with speeds of up to 45 knots and ranges of 3500 yards being achieved by around 1910. By this time a 21-inch torpedo had been produced and this enabled an increase in the size of the warhead-carrying capability. By 1916, the 21-inch warhead comprised 400lbs of amatol compared with 320lbs in the 18-inch weapon.

21-inch trainable torpedo tubes, 1909. (Collingwood Museum)

After the First World War, while the Royal Navy showed little interest in electric torpedoes because of the lack of speed performance achievable compared with compressed air drive, the Germans continued with electrical propulsion development and improved weapon performance up to a speed of 28 knots and 2000 yards range by 1918. For the Germans, the key tactical factors in favour of an electric drive system were the lack of exhaust gases, which made the weapon trackless, and the fact that the acoustic noise generated was much less than that for compressed air driven weapons. Neither of these attributes were fully appreciated by the Royal Navy until the Second World War when German weapons proved to be quieter and less visible than British and US weapons. Even more significantly, the superior noise characteristics of the electric weapon were eventually to usher in the era of the homing torpedo, which relied on relatively silent running to aid the acoustic detection of its target.

From the late nineteenth century, the torpedo was evolving as a very effective anti-ship weapon and when deployed in a submarine it presented a potent threat as the First World War approached. Initially, the limited range of both the weapons and submarines had little impact on naval warfare conduct in the open sea. However, as torpedo and submarine operating ranges increased, so the ocean-going threat to the British Grand Fleet increased. There was also an effect on the ranges at which naval surface engagements took place as capital ships stood farther off from torpedo-carrying destroyers.

One of the main contributors to increased torpedo firing range was the introduction of gyro stabilisation in the horizontal plane. This meant that, once launched and with the gyro running, the heading of the torpedo could be continually corrected onto the original straight line of the firing bearing until the propulsion system gave out. Any tendency to drift off the original target line, due to imbalance in torpedo mechanical characteristics, was reduced and effective ranges were increased accordingly. Thus, the introduction of gyro guidance technology in the 1890s started to drive up the ranges required from the guns of surface combatants in order to repel torpedo destroyers and, in turn, this demanded increases in gun calibre for more accurate fire at longer ranges.

Before the First World War, the short operating ranges of submarines had meant that capital ships on the open seas were generally unaffected by the torpedo threat and, arguably, this led to a complacent approach in the development of a counter to the submarine, a shortcoming which was to have a significant effect on the conduct of the Second World War at sea. Until the arrival

Torpedo firing from a V&W Class destroyer in 1918. (Naval Museum Portsmouth)

K Gun depth charge launcher during the First World War. (Museum of Naval Firepower)

of the depth charge, the only way of sinking a submarine was by ramming it before it had gone too deep. The concept of the depth charge had been mooted in 1911 as a 'dropping mine' but it was not until 1914 at the request of Sir George Callaghan,[28] Commander-in-Chief Grand Fleet, that the depth charge as a weapon went into production. The first weapons comprised steel drums filled with high explosive material. The drums were fitted with a pistol detonator which was actuated by water pressure and the setting of this operating pressure could be varied as a means of determining the depth of the eventual explosion. Each charge contained 300lbs of TNT which could be set to detonate at depths of up to 300 feet. Depth charges were not deployed by the Royal Navy until 1915. By 1916, Amatol[29] was being used as a more powerful and cheaper explosive and the pay load had been increased to 600lbs.

Initially, the depth charge was delivered by rolling it off the stern of the ship as originally proposed in the 'dropping mine' concept. However, the launch method soon developed into firing it overboard using a depth charge projector known as a K–gun, which could launch the weapons to port or starboard out to around 50 yards, in conjunction with other charges being dropped over the stern. In this way a pattern of some ten charges could be laid.

The chances of success were still slim because the submarine's pressure hull could stand the explosion, unless it was within some 20 feet. More significantly, there was still no way of tracking the submarine's position once it had dived except through hydrophones, which were capable of giving only a very approximate bearing but no range or depth information. Only nine German U–boats were believed to have been sunk by depth charges until 1918 when the first hydraphone equipment was fitted to ships and the final tally rose to 22.

General Electrical Systems

One of the first general electrical systems to be fitted in ships of the Fleet appeared in 1873 in response to the growing reputation of the Whitehead torpedo. It all came about when the Admiralty Torpedo Committee, established in 1870, was considering the options available for countering the night-time torpedo boat threat. One of the options considered was the use of electric lights to illuminate the area surrounding a ship in order to detect any attack. The Committee decided that it needed to see the concept demonstrated and, in 1874, the first electromagnetic induction

HMS *Inflexible*, 1876. (Naval Museum Portsmouth)

machine was installed in HMS *Comet* in order to power an illumination system. The machine used was a dynamo, manufactured by Henry Wilde of Manchester, and it was driven by a small steam engine rated at 6hp. This first motor generator set provided the electrical power for a carbon arc light, enclosed in a cylindrical case, and this generated an impressive 11,000 candle power beam for the illumination of enemy torpedo boats.

Although the illumination was restricted to the arc of the light beam, some 3–4°, and not all round as in the original requirement, the Committee declared 'The value of the light is decided and considerable.' By 1882, all Royal Navy ships were being fitted with what had become known as searchlights. They were classified as part of the ship's armament. Although by the late 1880s the searchlight's role had diminished, it was still a heavy user of electrical power and as such it dominated the design of ships' power generation systems until the beginning of the next century.

HMS Inflexible was launched at Portsmouth in 1876 using an electrical release system to break the traditional champagne bottle across her bows. Apart from her innovative launch and the largest rotating turret guns afloat, she was also fitted with underwater torpedo tubes and a new type of compound armour. A significant electrical advance was that she was the first ship to have electric lighting fitted during build. Despite this new feature, the governing electrical requirement was still the candle power needed from the two 'armament' searchlights which were fitted prior to

Searchlight onboard HM Torpedo Boat 74 in 1904. (Naval Museum Portsmouth)

her being commissioned in 1881. Accordingly, two 800 volt DC dynamos, manufactured by the Brush Electric Light Company, were installed to meet this requirement with the motor speed and carbon brush positions being adjusted until the required illuminating candle power was obtained.

The Brush dynamos were also used to power the internal lighting system which comprised small arc lamps throughout the ship. The circuitry consisted of two sets of 18 lamps arranged in series and powered from one of the dynamos. Initially, the failure of one lamp would extinguish the remaining lights but this design anomaly was later rectified by fitting each lamp with a mechanism to switch in a resistor to maintain continuity so that one failure did not lead to a blackout. Swan incandescent lamps were also put on trial in *Inflexible* but they were found to be unreliable. The Torpedo School was involved with the trials and, in addition to being concerned about the general safety of the high voltage generated by the Brush machine, the staff thought that the 800 volts contributed to the high lamp failure rates. The *Inflexible* installation was later modified to a 16 lamp system in four groups bringing the internal lighting operating voltage down to 200 volt DC.

Initially, both the ship's internal lighting and searchlights were fed from separate power generation and switchboard systems supplied by dedicated dynamos. However, a committee set up in 1882 recommended that a single electrical distribution system should be developed to allow the supply of electrical power for lighting and searchlights from any dynamo available on board; it was probably one of the first electrical damage control policy decisions. This led to dynamos either being connected in 'tension' – now better known as in 'series' – for high voltage requirements or 'quantity' –now in 'parallel' – for heavy current requirements. At the time, voltage and current measurement was only of academic interest as few methods of measuring the values existed outside laboratories. Indicative of this was that current was still being measured in webers and it was not until the Paris Conference of 1884 that the ampere officially replaced the weber and eventually became the standard unit of current under the MKS system of units introduced in 1901.

The safety hazard of working with 800 volts DC were brought to the fore less than a year after *Inflexible* was commissioned when, in June 1882, the first fatal electrical accident in the Fleet was recorded. The incident happened when a stoker grasped an un-insulated part of a lamp in a stokehold, to avoid falling due to the ship's motion, and was electrocuted due to an earth leak at the Brush dynamo. At the time the Society of Telegraph Engineers and Electricians was recommending that 200 volts DC and 60 volts AC should be the maximum voltages used for electrical equipment. It was following this incident that the Royal Navy decided to adopt 80 volts DC as a standard for on board power generation systems.

Although for the next 20 years the standard operational voltage was set at 80 volts, frequently there were voltage drops down to as low as 40 volts as the carbon rods, used in the arc lamps, burned away causing the loading effect to vary continuously. Ballast resistances were used to try and stabilise the dynamo outputs at 80 volts but eventually the solution was to discard the small

Searchlight remote control operator.
(Collingwood Museum)

arc lamp in favour of the incandescent lamp. However, the incandescent lamp was also rated in candle power and the applied voltage still had to be varied as necessary to produce the desired lighting level. Such was the sensitivity of the electrical system to voltage variation that, at one stage, it was recommended that any rating who switched off an electric light should inform the Officer of the Watch to ensure that the supply voltage would be adjusted!

An unacceptable level of lamp failures, caused by unintended operation at significantly higher voltages, led to a requirement to test each lamp to twice its working candle power before acceptance. The decision to use double the required lighting output as the test condition was alleged to have been based on the seamanlike practice of testing ropes to twice their working load, a clear sign of the fledgling nature of electrical technology at the time. By 1885, approximately 100 vessels in the Royal Navy had been fitted with searchlights but only ten were fitted with internal electric lighting.

It is of note that in 1883 an alternative lighting system using AC voltages was also developed. It was known as a 'secondary lighting system' because it used a high voltage AC supply transformed down to a working voltage in order to power the lighting system. A number of manufacturers produced machines which were capable of producing AC voltages, and in some cases both AC and DC voltages from the same machine, which could have been adapted to supply a lower AC voltage lighting system. However, transformer technology was very inefficient at the time and heavy power losses were experienced. This unresolved design problem, and the priority of retaining a DC supply for the carbon arc searchlight and improvements in DC dynamo commutator design, finally led to a decision in favour of retaining DC power generation systems as the standard for Royal Navy ships.

In the 1870s, ships' power cables were rubber insulated with a wrapping of cotton tape coated with preservative with the cables then being embedded in putty and run through the ship in teak casings. The rubber insulation was later replaced by a lead sheathing with layers of jute yarn instead of rubber. Neither of these cable designs was found to be satisfactory and, in 1888, a change to rubber insulation with a lead sheath proved to be much more successful. With some improvement to the rubber compound used and in the manufacturing process, this basic design of lead sheath and rubber insulation continued to be used up to the end of the Second World War.

During the 1880s, a typical first class ship would have had three dynamos installed, each giving 200 amps at 80 volts DC. One example was HMS *Canopus*, which in 1886 was fitted with 264 internal lamps and three searchlights, all supplied from a common 80 volt system based on three Gramme dynamos developed in 1869 by Zénobe Theophile Gramme, a Belgian industrialist.

HMS *Canopus*, a pre-*Dreadnought* battleship, 1897. (Naval Museum Portsmouth)

Two of the dynamos were sufficient to power the internal lighting, leaving one for powering the searchlights during peacetime cruising stations. When the ship went to general quarters for action, the internal lighting load was reduced to allow the two remaining dynamos to give priority to the powering of the searchlights. By the turn of the century the power demand in major ships had grown so much that 600 amp dynamos were deemed necessary. Typically, HMS *Powerful* and HMS *Terrible* – the largest ships in the world in the late 1890s – were fitted with 800 incandescent lamps and six 24-inch searchlights all fed from a single electrical system powered by three steam driven dynamos rated at 48kW.

In 1887, the first remote controlled elevating and training system for a searchlight was developed. It proved to be an operational success but was rejected for fitting in the Fleet as being too complicated and too costly. Also, by then the case for such a sophisticated searchlight was in decline. More importantly, the demonstrated system probably represented the first recorded use of an electric motor in a ship. Although this application was rejected for use in the Royal Navy Fleet, over the next ten years the use of the electric motor increased rapidly in other areas such as ammunition hoists, capstans, winches and boat hoists.

Apart from these operational systems, a major factor causing the spiralling increase in electrical load was the improvement in domestic facilities and living standards made possible through the use of electrically powered equipment. In 1898, electric fans were introduced for cooling and ventilation between decks and electric ovens and boilers were quickly seen as much safer alternatives to open-fire cooking facilities in the galley.

Above HMS *Terrible* in 1895. (Naval Museum Portsmouth)

Right HMS *Powerful*. (Naval Museum Portsmouth)

Portsmouth Dockyard took a lead in the design and manufacture of electrical systems and produced a successful 80 volt DC 400 amp dynamo. In 1892, the dockyard designed the 'Portsmouth Switchboard' and this became the first standardised naval electrical distribution system. It was fitted in HMS *Centurion* in the same year and had the facility for paralleling generators to allow electrical loads to be transferred between dynamos without having to switch off the equipment. In 1899, an electrical engineer was appointed to the dockyard to investigate solutions to ever increasing electrical loads and the problems of voltage drop. The investigation recommended increasing the voltage from 80 to 100 volts and in 1900 this became the standard voltage. This modest increase did little to satisfy the requirements of large ships and in 1903 it was decided to increase the standard voltage further to 220 volts DC. This significant increase was justified on the basis that tests had shown that '220 volt shocks were unpleasant but not dangerous.'

In 1902, emergency electrical supply arrangements were first fitted and in 1903 the Low Power Switchboard came into being and started to take over from batteries in the supply of firing circuits and electrical data transmission systems. Other notable events of the time were the introduction of bayonet lamp fittings and the provision of ceremonial lighting facilities, first used by HMS *Royal Sovereign* during a visit to Keil.

In 1908, HMS *Defence* was built as the first ship to have a 220 volt DC electrical system. She also made history in that a watertight electrical ring main system was installed for the first time in a Royal Navy warship. This was a fundamental departure from the simple central switchboard concept that had previously been used. In this early type of ring main, all electrical equipment was connected directly to the main power supply ring and, consequently, it also had to be installed such that flooding in the ship would not lead to short circuits and loss of power supplies to other parts of the vessel. Experience in the First World War showed that it was difficult to maintain the watertight integrity of equipment in a flooded compartment and that it was not easy to isolate short circuited sections of the ring main under the stress of action.

Around 1921, the problems of water ingress were resolved by the development of the first high power short circuit protection device. This was an unusual piece of apparatus called a 'mainguard' which protected the ring main in the event of the supplied equipment being flooded. The mainguard consisted of a miniature bell-shaped housing which contained a current-carrying, hollow carbon fibre tube connecting the ring main spur cable, entering the top of the bell housing, then vertically down to the load supply cable entering the bottom of the housing. Short-circuit

HMS *Defence* in 1907. (Collingwood Museum)

HMS *Ark Royal*, 1937. (Naval Museum Portsmouth)

protection was then achieved by the positive and negative spur cables off the DC ring main each being connected through a mainguard to the equipment being supplied. In the event of a flood in a compartment causing water ingress to the electrical equipment, a local flood switch was used to fire an explosive charge, adjacent to the carbon tube in the mainguard, causing the tube to shatter and thus isolate the ring main from the flooded equipment. The shape of the bell chamber was a unique feature in that it was designed so that as flood water rose inside it, the trapped air would be compressed forming an air pocket of such a size that the ring main connection would remain unaffected until the pressure head of flood water had risen to the point where the ring main installation as a whole would have been compromised by water damage.

A later development was to replace the mainguard with a fuse release switch whereby a toggle operated switch was held in the closed position against a spring by a stainless steel fuse. This fuse was then connected to the flood switch circuit which, in the presence of salt water, caused the fuse to blow and the switch to open and isolate the equipment. In the mid 1920s, the ring main system was redesigned and ring main breakers were introduced. In conjunction with the use of overload protection devices and low power relay control circuits, the new breakers provided automatic isolation of damaged or flooded sections of the ring main.

As the use of electricity continued to expand, particularly with the improvement in ship habitability, so the question of whether DC was the still the optimum power supply for newer ships arose again, especially as the US Navy had moved to using 440 volt AC main power supplies around 1900. The Royal Navy had first looked at using AC in 1883 and decided against it, partly because of the perceived importance at the time of the carbon arc searchlight. In 1932, in light of the US Navy's earlier decision, the situation was again reviewed by the Department of Electrical Engineering. The idea of a change was again dismissed on the grounds that the danger to personnel of 440 volts AC was too great and that the ratio of motor loads, the primary driver for using AC, to other loads was not high enough to warrant the changeover. At the time the electrical load in a typical cruiser was around 900kW and in a typical destroyer around 100kW, with the power being supplied by steam-driven turbo generators; at the time these machines were rated at around 225kW.

The 1930s saw a considerable increase in warship construction, including the introduction of a new class of aircraft carrier. HMS *Ark Royal* was the first ship of the class and she had six 400kW generators. The growing size of the electrical installations raised new doubts about the effectiveness of the fault protection system used on the ring main and also the ability of the circuit breakers to deal with magnitude of fault currents that could now be expected. In late 1938, HMS *London* was made available for extensive trials during which low impedance faults were imposed on the distribution system with various combinations of generators running in parallel. The maximum short-circuit current measured during the trials was 33,000 amps with the four generators in parallel and this was the maximum which could be interrupted safely by the type of ring main switch gear then in use. Discrimination between circuit breakers at these high current levels was found to be unsatisfactory and new over current relays had to be designed that could be fitted in existing ships. Circuit breakers had to be fitted with back-up protection in the form of high breaking capacity fuses, and switchgear makers were given the task of developing breakers of higher breaking capacity for future ship installations.

Communications

Following James Maxwell's prediction of the existence of electromagnetic waves in 1864, Heinrich Hertz succeeded not only in demonstrating the wave phenomenon but also measuring the velocity and wavelength in a series of experiments between 1885 and 1889. Given that the velocity was established as that of the speed of light, Hertz confirmed that the relationship between the wavelength and the frequency of the electromagnetic wave was such that the higher the frequency, the shorter the wavelength. Although not known at this time, the wavelength was also to have a significant impact on the propagation characteristics of the wave through the ether and the opportunities for naval warfare applications. In recognition of the Hertz experiments, which were seminal to the future of communications technology, the waves became known as Hertzian waves, or radio waves, with further recognition being accorded by the adoption of the Hertz as a unit of frequency, equivalent to one cycle per second, under the International System of Units (SI Units) in 1960.

The torpedo boat threat gave rise to the fitting of searchlights in ships in the 1870s and from then on Fleet trials and exercises were continually being conducted to develop ship defence measures. These trials not only included the tactical use of the searchlight but also the rigging

Torpedo Boat 2 in 1878. (Naval Museum Portsmouth)

Torpedo nets at work. (Collingwood Museum)

Wireless telegraphy equipment maintainer, 1898. (Collingwood Museum)

Captain Henry Jackson, founder of wireless telegraphy in the Royal Navy. (BAE Systems Archives)

of torpedo nets and hawsers between moored ships to prevent torpedo boats from manoeuvring through a stationary fleet. Exercises identified a major weakness in the use of torpedo boats as an attacking force at night in that they had great difficulty in recognising friendly ships which were in a darkened state, particularly when they were being dazzled by powerful searchlights. In 1890, a torpedo officer, Lieutenant Henry Jackson, suggested that Hertzian waves might be a way for a torpedo boat to indicate its position to friendly ships. The idea was not taken seriously by the Admiralty Signal Committee and, in any case, de facto responsibility for solving the torpedo problem was deemed to lie with the Torpedo Branch. In 1895, the now Commander Jackson was appointed to the *Defiance* Torpedo School where he found himself in charge of an electrical workshop facility and with the technical resources to carry out experiments using Hertzian wave concepts. Although unknown to Jackson at the time, Guglielmo Marconi had been working on similar experiments since 1894. When the two men met at the War Office in 1896, a comparison of notes showed that the principles they were using for their experimental apparatus were similar in that a spark transmitter was being used to generate the Hertzian waves and that a coherer[30] was being used to detect their presence. While there was a view that Jackson's apparatus was more robust, it was acknowledged at the time that Marconi's equipment was further advanced in terms of sensitivity and detection ranges.

The introduction to the 17th Annual Report of the Torpedo Schools dated 1896 tells that 'The Captain of HMS *Defiance* has been carrying out some interesting experiments of electrical signalling without conducting wires. The results generally are hopeful.'[31] The Captain referred to was Henry Jackson, who by now had been further promoted, and this first official account of an 'apparatus for electrical signalling without wires'[32] had been written by him. Perhaps unsurprisingly, the report also made reference to the fact that 'No system of signals by which Torpedo Boats and destroyers can challenge ships or make themselves known to friends has yet been devised and this should be considered by all officers as the want of some such system will greatly cripple their action in time of war.' While the report briefly mentioned Jackson's own work, it focused mainly on Marconi's apparatus which was described as being 'similar to those employed by *Defiance* but more fully developed'. Following the report, the Admiralty authorised Jackson to build two sets of equipment based on his experiments for ship trials.

Marconi trials transmitter, 1899. (Collingwood Museum)

The 1897 Annual Torpedo School Report confirmed that two sets of newly designated Wireless Telegraphy (WT) equipment had been built and used to carry out successful trials in the Hamoaze between *Defiance* and the gunboat HMS *Scourge*. During the trials, communications were achieved at a range of 5800 yards using a 6-inch spark transmitter and Morse code sent at a rate of eight words per minute. Later that year, Marconi achieved ranges of 24,000 yards in trials between ships using a 10-inch spark transmitter. The trials clearly showed the operational potential of the apparatus for communication in all weathers and over intervening land. In comments on the trial report, the concept of shore-based communications bases began to be recognised and even the idea of a portable system was postulated. The existence of radio interference was also observed and it was noted that 'if two or more ships try to signal at the same time, their messages are both recorded and probably both rendered unintelligible. This would necessitate similar rules as are at present in use for fog signalling.'

In 1898, a Commander Hornby witnessed Marconi's communication trials between Bournemouth and the Isle of Wight. During these trials Hornby noted that Marconi claimed to be able to 'attune' two transmitters and receivers by introducing a coil of wire containing an equal number of turns in the transmitting and receiving wires, or aerials. As such, Hornby reported that 'undoubtedly this method of signalling has great possibilities and its development by Marconi should be carefully watched, especially if his claim is substantiated of being able to confine his signals to certain instruments not absolutely in accord with them.' In other words, communications could be restricted between those ships tuned to the same frequency, unlike the earlier systems which effectively transmitted across all frequencies and could be detected by any receiving system. The concept of tuning transmitters to receivers was successfully demonstrated by Marconi on 28 February 1901 between Niton on the Isle of Wight and Poole. During the trials, communications between the two shore stations was achieved in the presence of untuned transmissions being made from HMS *Hector* at Portsmouth and HMS *Minotaur* at Portland.

Marconi tuned WT transmitter, 1900. (Collingwood Museum)

Marconi transmitter in HMS *Alexandra*, 1900.
(Collingwood Museum)

In 1898, Jackson left *Defiance* for HMS *Vernon* Torpedo School at Portsmouth, where he continued to develop a naval wireless transmission set. Subsequently, Marconi persuaded the Admiralty to carry out trials of his equipment during the peacetime Fleet manoeuvres of 1899 in the English Channel. Three Marconi sets were fitted for the trials to HM Ships *Europa*, *Alexandra* and *Juno*. Communications were established between the ships at ranges up to 100 miles with vertical wire aerial heights of up to 174 feet. The Admiralty was so impressed with the demonstration that three more Marconi sets were fitted to cruisers for Boer War blockade duties. At the time, Jackson was in HMS *Vulcan* carrying out transmission trials in the Mediterranean using equipment of naval design to investigate propagation phenomena. He used the results of these trials to present a paper on radio wave propagation in 1901 to the Royal Society.

In 1900, the Admiralty agreed to buy the Marconi apparatus providing it met a standard performance specification. This required that 'A ship in Portland and another at Portsmouth shall be able to communicate with one and other, each one having an aerial height of 162 feet, and each shall be able to communicate with a third ship having an aerial height of 100 feet to a minimum distance of 30 miles.' Marconi's designs met the requirement and later that year the Admiralty purchased 32 sets for use both afloat and ashore. Although the Admiralty had seen the potential of Marconi's invention and supported its development, surprisingly they did not oblige him to restrict its availability to Crown use. This led to the worldwide marketing of Marconi WT equipment and its adoption by many countries as a naval communications capability.

Marconi (left) with George Kemp and 10-inch spark transmitter and Morse Inker equipment, 1901. (BAE Systems Archives)

It was during the 1904–05 Russo-Japanese War that WT was first used in a major conflict and the Royal Navy, by virtue of the British alliance with Japan, was able to observe the potential of this technology at first hand at the Battle of Tsushima in May 1905. The Russian Baltic Fleet had been deployed some 18,000 miles from its home port round the Horn of Africa to relieve Port Arthur from the Japanese blockade. The Russian fleet WT equipment was of German design and had a tactically limiting communications range of only 65 miles, while the Japanese had Marconi equipment with greater transmission power and a much greater range. From the time that the Russians were sighted off the Philippines and then while crossing the China Sea, WT was used to report its position back to the Japanese Naval Headquarters. Eventually, Port Arthur fell before the Russians arrived and the fleet was diverted to Vladivostock on the Sea of Japan. Radio reports then kept the Japanese informed as to which of the three straits was being used by the Russians to enter the Sea of Japan. This knowledge enabled the Japanese fleet to position itself to 'cross the T' of the Russian fleet in the Tsushima Straits and virtually destroy it by the sinking of six battleships and three armoured cruisers. From a Royal Navy viewpoint, while WT played a significant tactical role in the Russian defeat, more importantly perhaps, the crushing victory brought about by the superior firepower and speed of the modern Japanese ships compared with the aging Russian vessels did a great deal to underpin the validity of Fisher's Dreadnought doctrine.

The maritime relationship with Japan led to a number of Dreadnoughts being built for the Imperial Japanese Navy in Britain. Typical of these was IJNS *Kongo*. The ship was a 27,500-ton battlecruiser built in 1911 and fitted out by the Marconi Company with the typical WT communications system of the time. It comprised a 25kW long range transmitter and a 1.5kW transmitter for short range use. The Double T Aerial, shown fitted on the *Kongo*, is of the cage type supported by spreaders 25 feet across at the top of the two masts; each of these were 180 feet high and had extensions fore and aft. The 25kW set was installed two decks below the main deck and was operated from below decks during wartime. In peacetime, there was a spacious operating cabin provided on the upper deck which also contained the 1.5kW set and an emergency motor generator with the latter being used to supply power in the event of a main power failure. The 25kW set used a standard Marconi synchronous spark transmitter with automatic frequency changing facilities and had a range of approximately 1000 miles.

By the time hostilities commenced with Germany in 1914, the majority of ships in the Royal Navy were equipped with one or two wireless sets operating on up to six tunes,[33] or frequencies in modern parlance. These frequencies were eventually grouped into bands which became associated with different radio wave propagation characteristics and, in turn, were used to classify the communications equipments which made operational use of those different characteristics.

Prior to 1938, the following internationally agreed band nomenclature was in use by the Royal Navy:

Imperial Japanese Navy ship *Kongo*, an armoured cruiser *c.*1911. (Collingwood Museum)

Low Frequency (LF)	<100kHz
Medium Frequency (MF)	100–1500kHz
Intermediate Frequency (IF)	1.5–6MHz
High Frequency (HF)	6–30MHz
Very High Frequency (VHF)	>30MHz

However by January 1938, it had become clear that the single, open ended VHF band was too restrictive to take account of the much higher frequencies required for use in both radio and emerging radar technology. Further international agreement led to the introduction of two more bands which would better cater for voice communications and, particularly, for the potential needs of radar.[34] These additional bands were designated as Ultra High Frequency (UHF) and Hyper or Super High Frequency (SHF). These bands will be discussed in detail later on.

In 1917, the Signal Branch had taken over the operation of wireless telegraphy from the Torpedo Branch, and the responsibility for research and development was transferred from *Vernon* to the RN Signal School in Portsmouth Dockyard.[35] It was around this time that alternatives to the Marconi and Jackson spark transmitters started to emerge. While the arc transmitter was initially considered as a possible replacement, with some examples still in service at the beginning of the Second World War, the main area of development was in thermionic valve transmitters. By March 1939, most equipment was based on valve technology but with transmitters and receivers having evolved almost independently and not as a communications system.

According to the size and role of the ship, transmitters could be fitted singly, in the case of a small ship, or as a modularised panel which could be configured with other panels to form a wireless set more suitable for larger warship communication requirements. Panel configurations varied across the Fleet with each set being given a different designation. In addition to a normal working mode, most warship sets offered low power and emergency operating modes and, occasionally, provision was made for incorporating VHF and voice transmission. In March 1939, the Royal Navy's Notes on Wireless Sets documented some 27 set options in service and employing various permutations of spark,[36] arc or valve transmitter technology; often each set had a different power supply requirement. Receivers would also be fitted singly or as part of a receiver outfit designed to cover the ship's operational frequency range. Receiver outfits were less prolific, probably because the design was not as sensitive to the ever increasing need for transmission performance, particularly in the areas of power and operating frequency.

Around 1937, it appears that the proliferation of unique systems of transmitter sets and receiver outfits was recognised as unsustainable. Apart from logistical complexities, each system had its own control facilities and power supplies, all of which carried their own training and management requirements. With this need to rationalise, the concept of the first integrated communications system, the Centralised Wireless System (CWS), came into being. CWS was intended to provide a flexible system which would improve communications control, transmission quality and reliability. To this end, the Central Communications Office (CCO) was introduced using standardised operating procedures and remotely operated transmit and receive facilities which were backed up by the local operation facilities in the transmitter and receiver rooms. Facilities for fast manual and auto high speed Morse operation were provided as well as increasing numbers of voice transmission channels. The latter was a particular requirement identified for carrier control of aircraft. With regard to ship services, AC power supplies were also selected as the standard in order to reduce interference problems inherent in the use of DC power supplies. It was not possible to retrofit CWS and only a few ships had the complete capability by the end of the Second World War.

It is worth recalling that Morse code was used primarily to pass information over ship communications circuits and that it was not until well into the Second World War that voice communications became more widely used by the Royal Navy.[37] The initial reluctance to use voice was due to the instability of the transmission frequency and the excessive bandwidth needed to accommodate voice modulation at HF frequencies. This bandwidth requirement was a limiting factor on the number of voice channels available in the HF band, particularly as frequency separation of the channel frequencies was also necessary to prevent interference between adjacent

A typical warship
wireless office in
1914. (Collingwood
Museum)

transmission channels. VHF technology offered a solution to the bandwidth problem and it had
been under continued development by the US for use in voice communications during the
late 1930s. Although the RAF had shown interest in using the VHF band for aircraft control
communications around this time, the acquisition of a full capability had not been given any
priority. Similarly, the Royal Navy had yet to see any real benefit of VHF for naval operational
application; this was possibly because of the relatively limited number of naval units involved at
any one time in an operation. On the occasions where voice communication was found to be
necessary, HF was considered to be operationally adequate.

Leading up to the Second World War, the Royal Navy had established a network of shore based
wireless stations which provided a regular communications service between the Admiralty and
the shore headquarters of the various Royal Navy Fleets around the world. These stations also
provided a service to individual ships at sea. HF frequencies were used for primary operational
communications circuits across this network because in this frequency band the transmissions
exhibited the propagation characteristic of sky wave 'bounce'[38] which under certain conditions
allowed global coverage. The shore stations were also able to transmit at LF frequencies and this
was used as a secondary circuit as the ground wave propagation characteristics in this band were
more reliable than sky wave[39] for obtaining global coverage. HF and particularly LF methods of
transmission and reception required massive aerial arrays and a great deal of electrical power in
order to maintain the integrity of the communications channel at a high level; this was a luxury
not available to ships.

Two other less well known areas of communications technology of the time were Sound
Telegraphy (ST), which is dealt with later under underwater systems, and internal ship
communications. Approaching the Second World War, the internal communications featured
alarm and ship address systems but, as yet, no internal telephone system.

Captain Henry Jackson rose to become First Sea Lord in 1915 by which time his passionate
belief in radio science had put the Royal Navy in a good position for waging the war at sea. This
involved not only the use of radio for communications but also for direction finding, a maritime
capability which was to have a profound impact on the future conduct of naval warfare.

Direction Finding

As early as the First World War, it had been observed that radio transmissions, initially introduced
to transmit information, could be used to identify the direction of the source from the position

of the receiver. With this capability in mind, during the tenure of Admiral Henry Jackson as First Sea Lord, a number of Direction Finding (DF) Stations were established ashore to intercept enemy wireless transmissions. These stations were able to work together and obtain positional fixes on enemy ships. During the war, they were used to detect the German High Seas Fleet radio transmissions and helped the British Grand Fleet to engage and contain the Germans after the Battle of Jutland in 1916.

While the tactical benefit of having a DF capability at sea had been recognised during the war, the development of a working system onboard ship proved to be problematical as there were complex differences in the operating environment at sea which manifested themselves as large and randomly variable errors in DF bearing accuracy. It is not necessary to present here the complete science behind the problems encountered, but suffice to say that it was not well understood in 1918 and only started to be clarified between the wars. However, for interest and understanding of events, it is worth summarising the problems encountered by the small research team at the Royal Navy Signal School that eventually found itself working under extreme pressure during the Second World War to deliver operational solutions.

Given that the intercepted transmission comprised an electro-magnetic wave, one source of major error was its interaction with the ship's own magnetic field,[40] errors being compounded by the latter continuously being re-orientated within the transmitted beam due to the elements and, more fundamentally, ship heading changes. Also, the ship in its entirety was radiated by the transmission and this meant that any reflections and re-radiation of the received wave from the superstructure and aerials would also be received by the DF aerial causing further distortion during measurement. These problems became more pronounced in their effect as the frequency of the transmission rose above 650kHz and the wavelength reduced below about 450m, thus becoming more comparable with ship length and, eventually, reducing to the length of deck mounted structures. At this point the transmission became more prone to reflection off structural surfaces and re-radiation effects were increased with more physical objects electrically resonating as if they were aerials. Adding to the problem, as frequencies increased above 1.5MHz[41] and into the HF band, under certain atmospheric conditions the sky wave component of the transmission increased. This component would arrive at an angle of incidence which would also distort the electro-magnetic field effect of the ground wave component of the transmission at the DF aerial resulting in continuously variable bearing errors. Finally, there was also the problem of bearing ambiguity. This arose because while it was possible to ascertain the bearing line of a transmission, using signal null or peak amplitude measurement, this measurement could not define whether the direction of the source indicated at the receiver was the true bearing or the reciprocal of that bearing.

By 1918, some progress on resolving the ship problems had been made with the development of methods for calibrating and correcting for the ship's magnetic field effect, which was the major source of bearing error at frequencies below 650kHz. By 1923, the eliminator reciprocal bearing had been invented and this resolved the ambiguity problem but again only up to 650kHz. With regard to reflection and re-radiation errors, it was decided that the only way to address this issue was for DF aerials to be sited clear of any other structure and, preferably, in a prime position at the top of a mast; this installation requirement was later to be in conflict with the wartime siting requirement for RDF equipment. This left the outstanding problems of poor performance above 650kHz and unacceptable performance above 1.5MHz. These were significant limitations as communications systems were increasingly using the HF band up to 30MHz and VHF technology was advancing rapidly.

During the 1920s, the Royal Navy Signal School tried various aerial systems to reduce error, including wire loops, frame coils built into the ship structure and rotating frame coils high on a mast. Trials indicated that the large wire aerials and fixed frame coils were not viable at HF and so research was focused on rotating frame coils, which were known to perform satisfactorily at MF and able to provide an MFDF capability for smaller classes of ship and submarine. Work continued to find an aerial solution to the HFDF problem and, in 1927, trials using a frame coil aerial mounted on the foremast of HMS *Queen Elizabeth* produced good results for up to 10MHz, but only in the presence of the ground wave component of the HF transmission. No real progress was made on resolving sky wave induced errors and ambiguity problems.

HMS *Queen Elizabeth* in 1924 before her HFDF trials fit. (Collingwood Museum)

By March 1939, the Signal School had an MFDF capability fully operational up to 650kHz and an HF/DF capability operational up to 20MHz but with limitations in that sky wave reception was unreliable and the reciprocal bearing problem remained. While this capability may have appeared limited for the Royal Navy, it was observed that no foreign warship at the 1937 Spithead Review had a DF aerial sited in a position where it could possibly have operated at frequencies above 1.5MHz.[42] In fact, post-war interviews showed that the Germans had concluded that HFDF was not possible and, as a result, they had stopped research and failed to appreciate the contribution that HFDF had made during the Battle of the Atlantic.

Radio Direction Finding

Hertz had observed that radio waves were reflected from metal surfaces in 1886. In 1904 based on this observation, Christian Hulsmeyer had been granted a British patent for this reflecting property to be used in a 'Telemobilescope'. This took the form of a radio wave transmission device for detecting objects and collision warning in a marine environment, but the device never proceeded to development. There is also evidence of advanced research work being carried out in this field by L.S. Alder at the Royal Naval Signal School. In 1928, Alder, along with a Captain James Salmond, submitted a patent application for a radio detection and location device based on the 'employment of the reflection, scattering, or re-radiation of wireless waves by objects as a means of detecting the presence of such objects'. However, it was not supported by the Admiralty Signals Department and work was stopped. In fact, the potential for military use was not of any great interest to the Royal Navy until around 1935 with the increased threat of war. Up until this time, radio wave transmission technology had been progressing in the commercial fields of radio

Sir Robert Watson Watt. (BAE Systems Archives)

and television transmissions but it was suddenly taken up by the military, became shrouded in secrecy and, for security purposes, was designated as Radio Direction Finding or RDF.[43]

RDF technology was based on the reflective properties of radio waves and what are now established as radar principles.[44] In the case of pre-war technology, this involved the transmission of HF radio wave energy causing any target caught in the transmission wave path to be 'illuminated' and to reflect part of that illuminating energy back to the original transmission source. Detection at the source allowed a target bearing to be established and, by timing the period between the transmission and the received reflection, the target range could be calculated. This reflective property was practically demonstrated in February 1935 by Robert Watson-Watt during the

Heyford bomber used in the Daventry experiment. (Fleet Air Arm Museum)

Watson-Watt's RDF receiver used in the 1935 Daventry radar experiment. (BAE Systems Archives)

'Daventry Experiment'. This demonstration involved a Heyford Bomber aircraft flying into the beam of the BBC's shortwave (6MHz) transmitter at Daventry and detecting the presence of reflected waves from the aircraft as an echo in a receiver 8 miles away; the receiver was designed by Watson-Watt. The success of this demonstration secured initial funding from the Air Ministry for radar development work at Bawdsey Manor in Suffolk.

Unfortunately, even in 1935 the Royal Navy did not think that radar capabilities were an urgent requirement. Naval resources were limited both by financial provisions, which had been largely vested in the RAF, and the Most Secret classification of the work, which meant that only the Signal School could be involved on the Royal Navy's behalf. These constraints meant that development progress was slow and focused mainly on producing RDF equipment designed to operate at a frequency of 75MHz.[45] This frequency had a wave length of 4m which was considered the maximum acceptable to ensure that the size of aerial would be practicable for fitting in ships,

The RAF RDF Research Establishment at Bawdsey Manor. (BAE Systems Archives)

HMS *Saltburn* minesweeper used for the first Royal Navy Type 79X RDF trials in 1937. (Naval Museum Portsmouth)

and then only major warships of cruiser size and above. In March 1936, the research department of the Signal School carried out its first shore-based trials of a naval radar system; working at 75MHz and manufactured entirely in house. Unfortunately, trials results for the RDF system were not impressive. Although it was able to detect an aircraft and measure its range, the system could not measure the target bearing or elevation. Furthermore, surface targets were not detectable at all.

Under pressure from the Controller of the Navy to identify naval applications for RDF technology, a 75MHz system was installed in a minesweeper, HMS *Saltburn*, and trials were first carried out at sea in December 1936. This system, designated Type 79X, used a simple aerial array strung between the masts of the ship. During the trial, aircraft were detected at a height of 5000 feet at a range of 17 miles but the system still gave no indication of target bearing. These results were again found disappointing and put down to the fact that the valve technology of the time was not capable of producing sufficient transmission power at 75MHz.

With valve limitations in mind, a decision was made to switch to a frequency of 40MHz and accept the resulting increase to a 7.5m wavelength and, consequently, an increased aerial size. This shift in operating frequency enabled the Signal School research to benefit from EMI developed valve technology which was then being used by the embryonic BBC television service. In addition, the RAF was already using this commercial technology for work being carried out by the Air Ministry at Bawdsey Manor on the development of a ground based aircraft detection system. Due to its land based nature, the RAF system was not constrained by aerial or equipment size and the development programme eventually led to the establishment of some twenty Chain Home RDF Shore Stations along the coast of England before the outbreak of war.

The Type 79X was installed in *Saltburn* and was successfully demonstrated as an air warning or surveillance system in early 1938. The results of the higher power, lower frequency Type 79X trials were promising enough for the Signal School to produce two more identical systems which were

HMS R*odney* prior to Type 79Y RDF fit in 1939. (Naval Museum Portsmouth)

designated Type 79Y and installed in HMS *Sheffield* and HMS *Rodney* within six months. Physically, the Type 79Y was a very rudimentary system and it relied on transmissions from one masthead aerial being received back at another aerial on a separate mast. The 7.5m wavelength gave rise to an aerial array which was around 4m square, a significant size at masthead heights. With these cumbersome array structures fitted on separate transmitting and receiving masts, each array then had to be manually rotated in synchronism from the radar office below decks. This rotation method was by no means elegant but it provided the principles for an effective aerial solution that enabled RDF technology to get to sea prior to the war and operate successfully beyond 1945.

Trials in *Sheffield* showed the Type 79Y to be capable of detecting aircraft at 10,000 feet out to ranges of 53 miles, reducing to 30 miles at 3000 feet. Results in the detection of surface ships were better than expected but the performance was unreliable. In any case, this was considered a bonus as a surface capability was deemed to be of a lower priority at the time. Confident following the trial results, the Admiralty ordered the first 40 production Type 79Y radar sets from industrial sources, all previous equipment having been built in utmost secrecy by the Signal School. The production sets were delivered for installation a matter of days after war was declared on 3 September 1939, at which time *Sheffield* and *Rodney* were the only radar fitted ships in the Fleet.

A modified version of the Type 79Y, designated Type 79Z and incorporating both mechanical and electrical improvements, was fitted for trials in HMS *Curlew* in August 1939. Although the Type 79Z variant retained the separate transmitting and receiving aerials, they could be tuned to any frequency between 39MHz and 42MHz and this available frequency spread was used to reduce mutual interference between ships at sea. The set also had more transmission power than its predecessor – 70kW versus 20kW – and was capable of detecting aircraft at ranges out to 120 miles at 30,000 feet and 70 miles at 10,000 feet. While detection performance was adequate in the Type 79Y, the bearing accuracy was no better than plus or minus 10° as the horizontal beam width was 60°. However, in the Type 79Z a facility was provided to measure the angular position of the half amplitude return signal while manually sweeping through the target. This process allowed the bearing to be measured within 5°. Type 79Z production sets started to be fitted in the Fleet from May 1940.

Other than promising detection and location performance, one of the most important outcomes of the early *Saltburn* trials was the discovery of lobes in the radio wave transmission coverage, caused by reflections from the earth's surface, and that these lobes sloped upwards thus reducing the illumination and chances of detection of low flying aircraft and surface targets. It was also established that the higher the frequency of transmission, the lower the sloping effect and with a consequential improvement in surface detection ranges. This first understanding of radar propagation characteristics caused the Signal School to adopt three strands of research based on operating frequencies suitable for long range air warning, range spotting for gunnery control and surface detection in that order of priority.

Accordingly, in its early research, apart from 40MHz, the Signal School had also been investigating the use of two other operational frequencies, 1300MHz (23cm wavelength) and 600MHz (50cm wavelength) but progress had been very slow due to the limited power output of the contemporary valve technology at these much higher frequencies. Although not a military priority in the mid 1930s, the benefits of achieving higher transmission frequencies were already recognised by the Signal School as offering significant reductions in aerial size and improvements in theoretical surface detection performance. The advantage of the reduced aerial size was that it would allow RDF to be fitted to smaller ships, which was to prove essential given the future convoy escort duties of Royal Navy destroyers and corvettes. Apart from reducing the physical size, higher frequency operation also offered the potential for narrower beam widths, more predictable propagation characteristics, improved range resolution and target discrimination in bearing; these attributes could be exploited for various naval warfare functions, including fire control and surface detection.

In 1938 with research in the doldrums and the dark clouds of war imminent, the decision was made to focus on the 600MHz transmission band for use in fire control systems and, perhaps more importantly, to consult with industry. In September 1938, GEC Research Laboratories were brought

New Street Chain Home RDF station. (BAE Systems Archives)

HMS *Curlew*, the RDF Type 79Z trials ship. (Naval Museum Portsmouth)

into the secret world of the RDF project and, by June 1939, a transmitter based on Signal School designs and GEC valve technology achieved a peak power output of 1.2kW at 600MHz. Thus work was able to start in earnest on the application of the technology to a new RDF system for gunfire control, designated the Type 282, with less than three months to go before the outbreak of war.

Submarines

After David Bushnell's invention of his manually propelled 'Turtle' submersible, which he unsuccessfully used to carry out the first underwater weapon attack in 1776, the development of the submarine stagnated as a satisfactory solution to the submerged propulsion problem was not forthcoming. The arrival of the steam engine did not do much for progress and it was not until J.P. Holland produced an electrically propelled submarine in the 1890s that a practical version of the concept was produced. Holland's propulsion system comprised battery driven motors with the batteries being recharged using petrol engines. The first HM Submarine was built to the Holland design and was launched at Barrow on 2 September 1901. By 1910, the petrol engine was replaced by the much safer and more powerful diesel engine and this formed the basis for future submarine propulsion systems until the arrival of nuclear power.

Despite the protests of the traditionalists, who saw the submarine as the carrier of underwater torpedoes and in the words of Admiral A.K. Wilson, 'underhand, unfair and damned un-English',[46] the Royal Navy's D Class submarines were successfully deployed by the beginning of the First World War. During and after the war, the Naval Staff tried to endow submarines with what now appear to have been eccentric capabilities, including aircraft launchers, howitzers and even smoke laying systems. Unfortunately, some of these ideas ended with disastrous results in terms of human life and operational effectiveness. By the time the Second World War approached, the submarine's offensive capability was mainly limited to torpedoes and a medium calibre gun for surface actions. It could also be used for covert operations.

A Holland Class submarine leaving Devonport. (Plymouth Naval Museum)

One of the main technical challenges presented by the submarine was its small size and this prevented much of the early electronics equipment, such as radio communications and in due course, radar, being fitted on board. The latter problem was not resolved until the Second World War was well underway and more progress had been made in the development of higher frequency, smaller equipment.

Underwater Detection Systems

The problem of combating the early submarine seems to have been of concern to only the few naval officers who realised the threat posed by its potential as a torpedo platform, given the rapid improvements in the capability of the weapon and the performance of the submarine itself from 1901. Thus it was that while the Royal Navy concentrated on building Dreadnought battleships in 1905, the Germans had started the build of U-1, the first U-boat in their fleet, in the same year.

The earliest thoughts on an electrical means of detecting a submarine were put forward by Lieutenant C.A. McEvoy in 1882 and are recorded in the Torpedo School Annual Report for that year. McEvoy's device was based on the principles of induction balance, whereby two wire coils suspended in the water were each connected in parallel with a similar coil at the surface. One set of designated primary coils was connected to a battery with the other, secondary set connected to a telephone speaker. Initially, the coils at the surface were moved relative to each other until a low noise was heard in the speaker, indicating a low level of induced current in the secondary coils. In the presence of a large metallic object below the surface, the inductive link between the primary and secondary submerged coils was increased causing the secondary current through the speaker to increase and, consequentially, increased the sound from the speaker. No further evidence has been found of this idea ever being followed up as a submarine detection system.

The next sign of interest shown by the Torpedo School in underwater detection appeared in 1893 when research into the potential use of underwater acoustics for navigation and signalling purposes was started. One of the first ideas investigated was based on a type of hydrophone system, designed by the now Captain C.A. McEvoy. The system comprised a platinum strip enclosed in a submerged, air-filled diving bell. The strip was connected to a battery at the surface. The operating principle was that underwater sound (pressure) waves felt through the diving bell caused the strip to vibrate and close the electrical circuit to the surface causing an alarm system to sound. McEvoy's system and various other designs of hydrophone, all based on microphones placed inside air filled containers, continued to be evaluated up until 1903, with little success. Eventually, the Captain of the Torpedo School reported to the Admiralty that hydrophone systems were not suitable for navigation or signal use. The reasons given were that they were too unreliable and, in any case, needed constant manning; this was a requirement which the Royal Navy was not prepared to accept at the time.

On 29 December 1903, the Commander-in-Chief Home Fleet wrote to the Admiralty proposing that ship defences should be evaluated during the Spring Fleet manoeuvres, with submarines being given the task of preventing ships from returning to anchor off Spithead. Methods of defence submitted for trial included lassoing the periscope and sliding a charge down the rope; towing a grapnel astern with a charge attached such that it could be detonated when in contact; and streaming an indicator net with a red flag attached to show when the submarine had been snagged. Apart from demonstrating that 'very much more successful and more reasonable methods' were required for ship defence, the trial showed that, in the presence of submarines, ships could not remain stationary and as a result their coal consumption increased.

In 1909, a committee was set up to look at available underwater signalling devices and, in 1911, two acoustic systems began to be installed in the Fleet. Both of these systems were commercially produced by the Submarine Signal Company, which was then based in the US and Britain. Both systems were found sufficiently capable of transmitting and receiving underwater sounds such that they could be used for sound telegraphy communication. However, one system could be more broadly tuned to permit reception of the sound of the bells used for underwater sound signalling by lightships as a navigation aid. It was this system which later proved to have a limited capability for the detection of the engine noise from a U-boat, and then only after modification by one Lieutenant Hervey of the Torpedo School.

In 1910, the Admiralty set up the Submarine Committee[47] and allocated HM Ships *Speedwell* and *Seagull* for the evaluation and development of anti-submarine tactics. A series of trials was carried out but the new ideas still mainly relied on the submarine being visually detected and, in some cases, its course estimated in order to be effective. Tactics explored included wire sweeps rigged between ships, machine gunning of periscopes, shelling of periscopes using delayed action fuzes and even the spreading of chemicals on the sea surface to effectively blind the submarine. It was reported that none of these methods were deemed successful except for the sweep system, which was the only one that offered an alternative, albeit limited, to visual detection. The possibility of sliding an explosive charge down the sweep on making contact was also tried as a means of sinking the submarine and this led to other methods of delivering a charge to be considered. Significantly, the Submarine Committee did not reopen the debate on the use of hydrophones or other acoustic means for detection but, as the result of its work, the real potential of the submarine threat started to be documented and the need for a detection solution became apparent.

Following the outbreak of war on 28 July 1914, Commander C.P. Ryan, an experienced WT Officer, was called out of retirement from working with Marconi and stationed in Scotland at Inchkeith. In the autumn of 1914, Ryan started to investigate the use of hydrophones for submarine detection by their engine noise and thanks to his perseverance and some support from Admiral Sir David Beatty, then Flag Officer in command of the Rosyth Battle Squadron, the Admiralty decided to support the research with a trials boat. This eventually led to the establishment of an Admiralty Experimental Station at Hawkcraig Point on the northern banks of the Firth of Forth in December 1915.

Ryan's experimental work in the detection of engine noise was based on the hydrophones produced by commercial companies such as the Submarine Signalling Company and the Automatic Telephone Manufacturing Company. His technique was to modify the commercial product, aided by his WT experience, and use practical trial-and-error to get the best results underwater. His modifications included the use of various forms of microphone enclosed in watertight housings and linked to a diaphragm that would respond to changes in water pressure generated by acoustic noise in the underwater environment. Initially, carbon granule microphones were used as the transducer to convert mechanical movement of the diaphragm into an electrical signal. The carbon microphone was later replaced by the moving coil microphone, which offered better frequency response characteristics and its transducing performance was less affected by the ship's attitude and motion.

This approach certainly identified, but did not quantify, the many problems of background noise in the water not only due to rough water conditions but also the noise of the detecting ship. However, by as early as February 1915, the Admiralty had already seen enough trials success to authorise the installation of what were omni-directional hydrophone systems to protect harbours around Britain from the submarine threat. These systems were intended to locate approaching submarines by triangulation of the source of the engine noise from intercepts made on a number of the hydrophones laid out on the sea bed. An extension of this concept was developed in 1916 and involved the use of hydrophones to monitor defensive mine fields and, on hearing the sound of engine noise passing through the field, the mines closest to the loudest source would be remotely activated. Both applications had their operational problems, not only with background noise but also when more than one ship or submarine was present in the area.

The Hawkcraig work also investigated the use of portable hydrophones on board ships. The early versions were portable adaptations of the hydrophones used in the fixed sea bed installations, with efforts made to tune to the bandwidth of 400Hz to 1000Hz. This bandwidth covered what was Ryan's estimate of the frequency band of a submarine's engine noise signature. In addition to optimising the equipment for the sound of the submarine, Ryan also felt from his experience that this bandwidth would exclude much of the higher frequency background noise to be found in the sea. Some 4500 of these Portable General Service (PGS) hydrophones were supplied

A portable directional hydrophone in use, 1917. (Imperial War Museum)

to fishing trawlers commandeered to support antisubmarine activity throughout the war. The PGS hydrophones were omni-directional and hung over the ship's side while stationary. They offered low probabilities of detection, even in calm seas, and could not be used while the ship was underway making them useless for submarine hunting.

In July 1915, the importance of the work being done by Ryan was further recognised when Their Lordships set up the BIR and one of the sections, Section 2, was given responsibility for the detection, location and destruction of enemy submarines. It was after the formation of this section that civilian scientific staff were seconded to work with Ryan in February 1916 at AES Hawkcraig. Following the involvement of BIR scientists in the development work at Hawkcraig, a more rigorous research programme was put in place, which sought to consolidate Ryan's empirical data with known acoustic theory, most notably that involving the properties of sound waves underwater and actual submarine noise signatures.

Unfortunately, this approach set in train some relationship difficulties between the BIR management and the Admiralty Operations Staff, where uniformed resentment grew as a result of the BIR's independence and the fact that the scientists seemed to be more interested in solving theoretical problems than delivering a working system to the operational coal face. Fortunately, these difficulties were felt less at the working level where the scientists were duly impressed by the progress made by Ryan under far from ideal conditions. Nevertheless, this friction led to a lack of cooperation and communications between the two groups, not helped by the fact that Fisher had been made the BIR Chairman. Fisher's uniformed opponents were reported to have referred to the BIR as the 'Board for Intrigue and Revenge' and continued to resent its independence and distrust its motives. It was under these conditions that underwater detection research progressed but in an uncoordinated way, which would not only have been internally disruptive but also must have been confusing to the French and American collaboration efforts going on in the field.[48]

The research approach used by Ryan at Hawkcraig to develop the hydrophone for detection was making virtually no progress in the determination of range, bearing or depth, key factors in the delivery of any anti-submarine weapons. This was clearly unsatisfactory and an independent review of the Hawkcraig work by the BIR scientists quickly concluded that an analysis of the sound spectrum, or noise signature, emanating from the submarine was needed and that the highest priority should be the development of a directional hydrophone.

In the same time frame, Ryan had started to develop his own form of hydrophone system for installation in submarines as a ship detection facility with a limited direction-finding capability. Typical of Ryan's practical approach, the system comprised port- and starboard-fitted hydrophones, tuned to low and high frequencies and baffled from each other by the mass of the submarine hull. From the received sounds, mostly high or mostly low, an experienced operator could assess the relative bearing of the source. Interestingly, in the initial development stages there was no way of predicting and testing the resonant frequency of hydrophone units. This resulted in them being selected empirically by, amongst others, one Lieutenant Harty RNVR.[49] He would tap the hydrophone face, listen carefully to the note produced, assess the peak resonant frequency of each unit and pair them as high or low for installation. This type of system started to be fitted throughout the Submarine Fleet in 1917. Some surface ships were also fitted with similar port and starboard hydrophone installations but these proved less successful owing to the noise of the ship when making way and the background noise at the surface of the sea.

At the end of 1916 and before the BIR work had progressed to the production of hardware, the decision was made to open a new Admiralty Experimental Station at Parkeston Quay, Harwich, the base port for the Harwich Force.[50] The civilian scientific team was then moved from Hawkcraig to the new station. Research at Parkestone Quay soon produced a number of inventions including the directional hydrophone and an acoustic ranging system based on a line of hydrophones being used to fix the position of an underwater explosion. The first trainable and directional hydrophone, the PDH Mk1 ([Portable Directional Hydrophone Mark 1) was produced in May 1917.

Indicative of the problems between the Navy and the BIR, Ryan, who was still at Hawkcraig, adapted some of the scientists' sound-baffling techniques from the PDH Mk1 in his own version of the system, designated PDH Mk2. Neither system exhibited the full operational performance characteristics required but later in 1917, both PDH Mk1 and Mk2 equipments started to be supplied to ships; mostly smaller units such as fast motor launches and trawlers. Unfortunately, poor detection performance when underway and the few depth charges carried by these boats meant that anti-submarine success was limited. In addition, the German U-boats quickly learned the limitations of the hydrophone and rigged for silent running, which involved slow speeds and silence in the boat. The result was that, throughout the war, only four U-Boats are thought to have been sunk as a direct result of hydrophone detection.

Although the deployment of hydrophones in submarines had some operational success, surface and own ship noise continued to be problematical for ship-installed systems and this gave rise to the idea of the towed hydrophone. First investigated in America and nicknamed the 'Rat' because of its rodent-like appearance, the equipment comprised a number of omni-directional hydrophones encapsulated in rubber. The Rat was towed astern of the ship and at a depth intended to minimise surface noise. Under the British code name of 'Chunk', ASD took this idea and designed a variety of directional versions of the towed hydrophone, which went by names such the Nash Fish, Lancashire Fish and Porpoise. Although the principle of the towed hydrophone appeared sound, a distinguished BIR scientist working in the field at the time, Dr A.B. Wood, was unimpressed and observed that 'At speeds above three or four knots, extraneous noise due to towing rope vibration, water noise and own ship's propeller noise tended to mask the "wanted" noise of a distant ship or submerged submarine and ranges of detection were consequently rather small.'

Post-war, hydrophone detection systems continued to be developed for submarine fitting and, by 1922, the Type 700 set offered a much improved detection and direction-finding capability. The equipment comprised three fixed hydrophones, sited at port, starboard and astern positions in the submarine's hull, and a directional hydrophone with outputs that could be monitored directly by the Captain.

Although the use of hydrophones for detection had been at the forefront of research during the war, the possibility of using them for sound telegraphy, or underwater communication, had

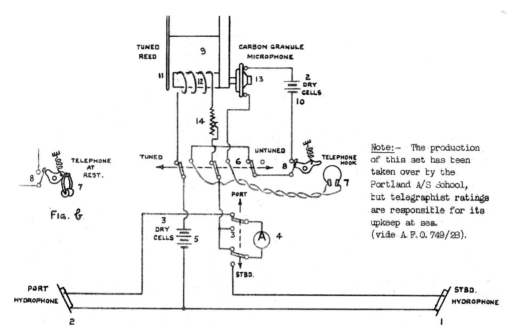

Type 706 sound telegraphy equipment from 1928. Note the reference to WT Branch maintenance responsibility at sea. (Collingwood Museum)

been investigated, with little success, since the 1890s. One of the earliest operational sets was the Type 102. Developed in 1916 by the Torpedo School, it used two sonic oscillators, invented by Professor R.A. Fessenden in 1913,[51] to mechanically resonate a transducer at 540Hz causing pressure waves in the water which could be modulated using a Morse key. This finally made underwater communications with submarines possible. A transmit-only set, Type 103, was also produced for one-way communication to a submarine being towed by a surface unit; standard practice for extending its operating range at the time. The justification for the Type 103 was to allow secret underwater communications with the submarine when slipping it from the tow in the presence of a detected enemy threat.

Other lines of research which demonstrate the lasting contributions to underwater technology made at Parkeston Quay by the BIR scientific team included the magnetic mine,[52] highly innovative mine detonation systems for use in harbour defences, cavitation masking and asdic technology.

The idea of generating sound waves underwater had been around since 1912 when L.F. Richardson had suggested, at the time of the sinking of the *Titanic*, that ultrasonic sound waves might be used to locate underwater objects by echo. In 1916, Paul Langevin discovered that a property of mica crystal, when connected to an electrical voltage, was that it vibrated at around 100kHz. This property, known as the piezo-electric effect, had been discovered by Pierre and Jacques Curie in 1880 using quartz. When this vibration was induced underwater it would generate travelling sound pressure waves in the water.

It was the principles demonstrated by Langevin that formed the basis for the BIR research done at Parkeston Quay in a new technology field given the codename ASDICS.[53] The acronym apparently came from a contraction of Anti-submarine Division followed by 'ics' to bring it into line with other disciplines such as physics and mathematics. The British research programme selected quartz as the material for a transducer which, when electrically excited, could convert the electrical energy into mechanical energy and send a pulse of pressure waves through the water. When these waves hit against a solid object, such as a submarine, then a sound pressure wave would be reflected back towards the transducer as an echo, where it could be sensed as a vibration and be converted back into electrical energy. By measuring the time between transmission and reception and using the known speed of sound in water, the range of the object could be calculated. The point at which the reflected wave was at a maximum in intensity gave a measure of the bearing of the object.

Experimental trials were carried out using a transducer which was mounted outboard of the ship and could be trained through 360° and tilted downwards. A major breakthrough came in March 1918, when asdic echoes were obtained from a submarine at a range of 500 yards. The outboard system was not much practical use for anti-submarine warfare and work quickly moved on to an inboard version.

The first inboard asdic system was fitted on HM Trawler *Ebro II*, a Parkeston Quay trials vessel, in October 1918. However, the work was not completed until 16 November 1918, five days after the Armistice was signed. This prototype, later designated as the Type 111, used the piezoelectric properties of quartz to produce acoustic energy waves in the water at operating frequencies ranging from around 20–50kHz, depending on the applied excitation voltage. Valve amplifier designs, based on existing WT technology, were used to provide the required excitation voltages, anywhere between 6000–10,000 volts. The receiver amplifier used to detect the low levels of returning energy was battery powered.

Efficient transfer of both transmitted and received acoustic energy was achieved by installing the transducer in a free flooding, retractable canvas housing, or dome. The dome was cylindrical in shape and, when protruded outside the line of the hull, it provided a still water environment at all speeds. The transducer directing gear was movable in training for submarine search and hinged to allow rotation from the vertical to the horizontal for use as an echo sounder. During trials, the Type 111 was found capable of picking up returning echoes out to 600 yards. The Admiralty had anticipated the success of the Type 111 trials and pre-ordered 22 sets in June 1918, but these arrived too late to affect the outcome of the wartime anti-submarine battle.

In November 1918, the signing of the Armistice brought about a rapid reduction in the funding available for wartime research and this slowed down both asdic development and ship fitting programmes. Eventually, the first production equipment, Type 112, was fitted in early 1919 to a patrol boat, P59, and, in due course, other P Class vessels which then formed part of the Portland Anti-submarine Training Squadron.[54] The Type 112 was a production version of the Type 111. It was simply engineered with training and stabilisation of the transducer against pitch and roll being carried out manually by the operator and the transducer bearing being transmitted to the ship's bridge using rod gearing. Range was obtained by the operator using a Morse key to initiate the acoustic transmission, simultaneously starting a stopwatch and then listening for a returning echo in his headset. The set could be used for either signalling or echo detection with communications being achieved at ranges up to 10,000 yards and returning echoes being recorded on targets out to 3000 yards at speeds of around 15 knots. Overall, the Type 112 represented a significant increase in capability over portable hydrophone systems. One shortcoming was that the acoustic beam width was only 5° and this was restrictive during convoy screening and search operations.[55] Later the Type 112 was adapted for fitting to destroyers as the Type 114 and put into service with the 6th Destroyer Flotilla.

It was not until 1922, that the first submarine, *H-32*, was fitted with an asdic system. The Type 113 was a variant of the Type 112 and similarly fitted in a retractable dome that emerged from the top of the submarine casing. This arrangement meant that while operations on the surface were not possible, the set could be operated when submerged and, more importantly, when bottomed. The transmission frequency was in the range of 20–21 kHz.

The narrow 5° beam width of the Type 112 quickly proved to be a significant tactical limitation and, in 1923, attempts were made to improve search capability by incorporating a 180° wide beam, switchable to 5°, in a Wide Angle Sweep (WAS) set, designated the Type 115. This set was initially fitted with two of the standard 15-inch quartz steel transducers. One was mounted horizontally above a cone-shaped sound deflector to provide a 360° surveillance capability and the other was mounted vertically to be conventionally used for submarine hunting. Four Type 115 sets were fitted to R Class destroyers in 1925 but results proved unsatisfactory in wide angle mode despite replacing the single transducer with two 5 -inch transducers each covering a surveillance arc of

P Class patrol boat, the first class to be fitted with the hull-mounted Type 112. (Collingwood Museum)

H Class submarine with Type 113 asdic dome and torpedo launch facility. (Collingwood Museum)

180°. Later in 1925, work was abandoned on the Type 115 and became focused on improving the 5° variant. This decision was due to have a profound impact on the development of anti-submarine warfare tactics. Mainly, this was because the Royal Navy entered the Second World War with a restricted underwater surveillance capability for use in used in convoy screening operations against U-boats with much improved seagoing performance, particularly diving depth, since the earlier conflict.

Between the wars, the shipboard engineering of the asdic system was gradually improved and two of these improvements were featured in the short-lived Type 115 design. These were the introduction of electrically powered transducer training and streamlined domes fixed to the hull to ensure continuity of the water as a transmission medium at the dome interface at even greater operating speeds.

In 1927, asdic research was moved from the Signal School to HMS *Osprey* at Portland and finally co-located with the Anti-submarine Training Squadron based there. The Portland work was supported by the Admiralty Research Laboratory at Teddington. While the move to *Osprey* meant that the nature of research became more closely linked to the solution of equipment and tactical problems, little progress was made in understanding the underwater acoustic environment. It was only when two sets of acoustic trials conducted in the Mediterranean at different times of year produced significantly different results that things started to change. In 1933, it was finally recognised by *Osprey* that more knowledge was required about the behaviour of sound waves through the water if asdic was to realise its full potential. From this point onwards, factors such as water salinity, water temperature,[56] pressure variation, reverberations, sound attenuation and own ship noise were given a higher priority for research investigation in order to explain the apparent day-to-day variation in asdic performance, which was casting doubts on the operational effectiveness of the Royal Navy's capability in the field of anti-submarine warfare.

Features developed in later asdic sets included more accurate data transmission systems, gyro stabilisation of the transducer bearing to compensate for changes in ship's course during operation and built in procedural training facilities for the operator. One significant advance was the development of a chemical range recorder that gave a permanent record of target range being plotted against time. This information provided an indication of target range rate which could be used in computing the submarine's future position. A final touch for improving command information was the introduction of the ubiquitous bridge loudspeaker, as featured in countless war films such as *The Cruel Sea*.

The majority of these improvements were embodied in Asdic Type 119, the prototype of which was fitted in HMS *Torrid*, a First World War R Class destroyer, around 1929. The production sets were first fitted in the B Class destroyers, which were the second new class of destroyers approved for build following the naval force reductions internationally agreed under the Washington Treaty in 1925. Type 119 eventually saw Second World War service, albeit by then heavily modified, and formed the basis for most asdic set designs up to the beginning of the war.

It was this slow but steady progress in the asdic 'system package' that may have led to some complacency in the newly formed Anti-submarine Branch and the lack of recognition of shortcomings in the Royal Navy's submarine kill capability. In addition to the beam width constraints mentioned earlier, neither the crucial issue of establishing submarine depth nor the limitations of the depth charge, both of which affected kill probability, appeared to have been given any high priority. This serious omission could well be attributed to tribal boundaries brought about by the fact that the Signals Branch retained responsibility for asdic development and the Torpedo Branch were concerned with depth charges; the only ship-fitted anti-submarine weapon at the time. This division of responsibilities left the Anti-submarine Branch in the middle with warfare tactics development responsibility but little control over improving equipment capability to support those tactics.

One example of a tactical shortfall was the loss of asdic contact information, which inevitably took place when a ship overran a submarine in order to launch depth charges over the stern. This limitation came with the weapons and tactical concerns eventually led to the consideration of launching the weapons ahead of the ship. While this concept had been briefly addressed in the mid 1920s, it was not followed through partly because of financial constraints and partly because of the lack of an asdic range recording facility. Although the latter appeared in the early 1930s, it was not until immediately before the war that the ahead thrown weapon was reconsidered.

The Royal Navy experience gained against U-boats during the Spanish Civil War in 1937 served only to further expose the limitations of its asdic equipment. This experience gave the lie to an upbeat statement regarding anti-submarine warfare capability made by the Admiralty in response to the 1936 announcement by Germany about its proposed submarine warfare doctrine. The initial phases of the Battle of the Atlantic were only to highlight the Royal Navy's shortcomings against a threat which had increased in potency without really being noticed.

At the outbreak of war, there were some 200 asdic-fitted warships and minor vessels available to the Royal Navy. The most modern ASW tactical set was the Type 128 fitted to destroyers with a variant, Type 127, being fitted to sloops. The Type 128 had a retracting, streamlined dome, a gyro-stabilised transducer, range and bearing recorders but with similar signal processing electronics to the earlier Type 119. Significantly from an operational point of view, the set had no depth measuring capability.

The Electrical Gyroscope

The mechanical gyro had been discovered in the early 1800s and, as referred to earlier, it found a use in the horizontal stabilisation of torpedoes. In 1885, a Dutchman, Marinus Van den Bos, patented the first gyro compass. However, the device never worked properly and it was not until 1890 that G.M. Hopkins invented the first electrical gyroscope. Further development of this invention brought about the solution to two major ongoing problems that were being experienced by the world's navies. The first problem was the unreliability of magnetic compass systems for navigation purposes when installed in steel-hulled ships and when used at higher latitudes. The second was the increasing number of electrical machines being fitted and the electro-magnetic fields being set up by such machines causing distortion of the earth's magnetic field effect on any adjacent magnetic compass. In 1906, the interference problem of the growing number of electrical machines was minimised in HMS *Dreadnought* by fitting a master magnetic compass sited away from any electrical machinery and using a newly invented angle data transmission system known as a Bolometer to pass the gyro information around the ship.[57]

In 1910, the Anschutz Gyroscope was developed and fitted as standard equipment in Royal Navy battleships and cruisers before being replaced in 1912 by the Sperry Gyroscope. Not only did the introduction of the gyro give the Royal Navy greatly improved warship navigation capability but, in 1915, Sperry used the gyro to provide the sensing device for a ship's stabilisation system. The gyro was soon being adapted to provide stabilisation against ship's motion for the ship's weapon systems, some twenty years after its first application in the guidance of torpedoes.

During the 1920s, the accuracy of the gyro compass was complemented by the development of a Ship's Course Transmission System which allowed electrical transmission of reliable compass heading information throughout the ship. Later in the twentieth century, the principles of gyro technology were to become fundamental to the stabilisation of sensor, torpedo, gun and missile systems for the generation of accurate fire control solutions. Another major contribution was to come during the Second World War when gyro transmissions were used to north stabilise the presentation of radar information against changes in ship's heading. This proved to be an important innovation – the beginnings of a revolution in naval warfare management.

CHAPTER 3

MANAGING TECHNOLOGY
800–1938

Manning in Medieval Times

England had been fending off Viking invaders since the eighth century and in their national defence the English regularly had to raise armies at numerous locations around the coast as far apart as Anglesey, Northumbria and the Isle of Wight. The cost of maintaining a standing army to deal with such a widespread problem was unsustainable and so during the ninth century the English kings had established their right to recruit men to fight for the homeland under the concept of the Fyrd. This was the name given to the Saxon militia which could be called up as necessary to support the army. It involved an obligation being placed on communities around the country to provide a number of men, complete with arms, for the King's service in proportion to the size of each community.

While the Fyrd was initially a land army, by the eleventh century this obligation had been extended to cover the provision of men for service at sea. There is some recorded history of selection for sea service and it would clearly have been sensible to take men who made their living on the sea. However, landsmen were not excluded and, particularly if they lived in the coastal areas, they were liable to be used to make up numbers when necessary. For economic reasons, the reigning monarchs retained very few royal ships in their permanent service, preferring to commandeer merchant ships and the crews, paying them off when they were no longer needed. At the time when the primary purpose of having a fleet was to transport the army, this was a workable arrangement. However, as the nature of conflict evolved from the movement of troops into a more seafaring and offensive nature leading up the Hundred Years War, so more merchant ships had to be commandeered into the King's service with no certain knowledge as to when, or if, they would be returned to their owner.

Although Edward the Confessor appears to have started the process of commandeering ships in the eleventh century under an arrangement with the towns which later formed the Cinque Ports (Dover, Hastings, Hythe, Romney and Sandwich), it was not formally established under Royal Charter until the twelfth century when Henry I inserted an obligation to provide ships for the King's service into the charters of towns around the coast in return for certain privileges; notable amongst these towns were the five Cinque Ports in the south of England. These privileges included exemption from taxes; rights to administer local government and justice; rights to levy tolls; and to take ownership of salvage washed ashore in the port's vicinity. Under the obligation, the provision of ships was not just for campaign service but also for more routine matters, including the transport of the royal entourage to the continent. Subsequent monarchs revised the terms of the Royal Charter with the last changes being made by Charles II in 1668.

One reason for choosing the Cinque Ports was the constant supply of vessels undertaking commercial activities along the coast and the availability of the crews who served in those vessels. Another pragmatic reason was that, with the demise of the Viking threat, these ports were also the places at greatest risk of attack from the continent. This gave the citizens a vested interest in the King's naval activities not only for self preservation but also, in the case of English foraying expeditions over to the continent, for the promise of a share of any spoils arising from the venture.

The rest of the country was not excluded from the obligation for what became referred to as the Ship Fyrd and, if the entire fleet could not be supplied by the Cinque Ports, then more inland communities were charged with financing the cost of, rather than actually providing, a ship in defence of the realm.

Warship Manning Structures

Under the Ship Fyrd, manning of the King's ships was clearly divided into the ship's naval crew supplied by the owner and the soldiers carried for fighting. The ship had a civilian naval master, whose main responsibility was sailing and navigating the ship, and he was usually supported by a constable, or head officer, and a clerk. The master had no jurisdiction over the soldiers carried on board ship as they had their own chain of command which would have been headed by a delegated military commander who had been granted a king's commission or, possibly, the King himself.

Following the discovery of the New World – with its potential opportunities for trade and expansion – and the Reformation, during which relationships with the catholic powers of France and Spain deteriorated, the country's requirement for a more substantial standing naval force rapidly increased. In response, Henry VIII brought more royal ships into permanent service and so ship manning structures started to develop under a ship's captain who had responsibility for his ship and all fighting, sailing and logistical matters associated with it.

Broadly speaking, in a royal ship, the captain would have had a departmental structure to address fundamental areas of responsibility on board. In a larger vessel, this structure would typically include a first officer, a sailing master, a purser, a surgeon and an officer in charge (OIC) of marines (later Royal Marines) when these seagoing soldiers were carried on board.

Commissioned Officers

In a warship, only the captain, officers of lieutenant rank and above would have held a royal commission issued by the Lord High Admiral, later the Admiralty Board. Such a commission gave these officers military authority over all personnel in the ship in the name of the monarch. Having been given this authority, all sea officers, as they were generally known, came to form the Military Branch[58] of the Royal Navy and it was only from this group that officers could be selected for ship command.

The First Officer, second in command and later known as the First Lieutenant, would have had executive responsibility for managing the conduct of ship's company and naval warfare operations, including ship's gunnery. The junior lieutenants in the first lieutenant's department carried out delegated ship fighting duties and, as part of the Military Branch, in due course they could be promoted and given command, either at sea as a ship captain or in a naval shore posting.

When carried, marine officers had the commissioned authority of the Military Branch in their own sphere of operations. Thus the OIC would have had specific responsibility for the fighting effectiveness of the marines and their day to day ship security duties, such as guarding the captain's quarters or the spirit store.

The naval wardroom was formally introduced in 1745 as the place in ships where all commissioned officers took their meals. Initially, its use by other officers was dependent on the size of the ship but, in principle, only Military Branch officers had full rights of access, with other professional heads of departments, such as the master, surgeon, purser and chaplain, being admitted more routinely where separate accommodation was not available due to the size of the vessel.

Warrant Officers

Warrant officers were so called because they were appointed by the Navy Board[59] to ships in accordance with the manning warrant, later referred to as the ship's Scheme of Complement. As a group, they represented both the Military Branch specialists and the non-Military Branch specialists, the latter being collectively grouped as the Civil Branch.

Military Branch warrant officers included the gunner, boatswain and carpenter who could, if circumstances allowed, achieve a commission. The artisans, or tradesmen, amongst these warrant officers were also regarded as the ship's standing officers as they remained with the ship when it was placed in 'Ordinary'.[60] Until the mid nineteenth century, the warrant officer specialists who made up the Civil Branch included the sailing master, purser, surgeon, instructor and chaplain.

The master was usually the senior warrant officer on board and responsible for sailing matters including navigation and the manoeuvring of the ship. The importance of his role was recognised by the fact that he usually messed in the wardroom with the Military Branch officers. This happened particularly in smaller ships where the post existed in the ship's Warrant and the sailing duties were not undertaken directly by the captain. The promotion route to becoming a master was open to anyone who felt that they had little chance of gaining a Military Branch commission. As such, it was used by midshipmen, who could not pass the promotion board for lieutenant, as well as by ambitious lower deck seamen to gain advancement to warrant rank. From 1843 masters were granted commissioned status and with this came access to the wardroom.

The professions of surgeon, purser and chaplain were represented at warrant officer level with the surgeon usually being invited into the wardroom. Although chaplains were often considered as a luxury in ships below third rate, they were also granted commissions in 1843. Artisans such as sailmakers, ropemakers, armourers and coopers could achieve warrant status in larger ships, but in smaller ships they were more likely to be rated petty officer.

The Military and Civil Branches of the eighteenth century formed the structural basis of the Royal Navy with significant differences in conditions of service and with the clear differential that only Military Branch officers could achieve to command. One other obvious difference was that Military and Civil Branch officers lived in separate messes on board the larger ships with the Military Branch officers of lieutenant rank and above in the wardroom mess and all others in the gunroom mess. In 1875, all Civil Branch officers of lieutenant rank and above were finally given full wardroom status. This then meant that in ships which still had a gunroom, all officers below the rank of lieutenant messed together and this remained the case until the demise of the Capital ship after the Second World War. Military and Civil Branch officers were also listed separately in the Navy List, first produced in 1782 as Steel's Navy List and from 1814 as the Navy List. This method of documenting the officers serving in the Royal Navy eventually generated the terms 'Military List' and 'Civil List', reflecting the two main branches, which were to survive into the twentieth century.

The Civil Branch started to evolve most significantly in the nineteenth century with the arrival of the engineer and the electrician. When the engineer appeared in the Fleet around 1830, he was initially categorised as being an artisan and, generally, had petty officer status. This was changed to warrant officer status in 1837 when the engineer artificer was introduced and, finally, commissioned status in 1847. This final acceptance of the commissioned engineer officer signalled the beginning of the end of sailing technology as a primary contributor to the sea battle and the boundaries between the Military and Civil Branches started to be challenged.

In 1902 when the Selborne-Fisher officer training scheme was introduced, engineers who eventually qualified under the scheme were included in the Military List. However, those engineers who preceded the scheme remained on the Civil List until 1915. For reasons covered later in this book, it was not until 1925 that the use of the term 'military' was finally withdrawn and separate Executive and Engineering Branches were introduced.

It is of interest that the composition of the Military Branch itself was also subject to evolution. In the glory days of sail, the carpenter had been in the Military Branch but, in 1878, he was re-categorised into the Civil Branch when the demise of the sailing Ship of the Line had been finally accepted. The carpenter, as a specialist artisan branch, continued until February 1918 when it was re-titled as the Shipwright Branch.

Military Branch Training

For much of the time since its inception, the Royal Navy had carried out the initial training of its Military Branch officers on board ship. Potential officers were usually recruited from upper and middle-class families under a process known as 'nomination'. Typically these recruits were the sons of noblemen, landed gentry or wealthy merchants. The nomination process involved parents putting forward their sons to be accepted on the books of a King's ship under what was known as the 'Captain's Servant Entry'. This form of entry involved a ship's captain, often a friend of the family or with at least some social connection, undertaking to supervise the young man and train him for service in the Royal Navy. Nomination was made possible by an official ruling that senior officers were authorised to have such 'servants' on the ship's books and receive payment from the Crown.

Unfortunately, the Admiralty had no real control over this method of recruitment and it was possible to put the name of a boy down for a ship at birth. Although such boys could not generally be sent to sea until they had reached a minimum of eleven years of age, the time on a ship's books did count towards the service time in the Royal Navy required before becoming eligible for promotion to lieutenant.

After four years of satisfactory service in the ship, the young boys would normally be made up to midshipman and undergo an arbitrary level of training, depending on the interest of the captain and any instructor tuition available on board, before promotion to lieutenant. This promotion to commissioned officer status was frequently dependent as much on patronage as on performance. While the nomination process served a recruiting purpose, it did not weed out unsuitable candidates. It was also open to abuse, with not infrequent courts martial of captains for drawing the pay on behalf of notional boys carried on their books and themselves pocketing the remuneration offered by the Admiralty.

In 1676, the Secretary to the Navy, Samuel Pepys, introduced a more controlled method of recruitment whereby the Admiralty nominated suitable sons of gentlemen under a scheme officially known as 'Volunteers per Order'; these 'Volunteers' colloquially became known as 'King's Letter Boys'. In addition, Pepys also introduced an examination board for promotion to lieutenant. This involved the candidate being verbally examined by a board of senior officers which is little different, in principle, to the Midshipman's Board used in the twentieth century to justify promotion to acting sub lieutenant. The laudable intention was to prevent social connections, rather than ability, being the key to promotion. Unfortunately, even this process was not infallible, with Boards still being able to make decisions based on personal or social prejudice.

Royal Naval Academy Portsmouth, 1880. (Naval Museum Portsmouth)

The First HMS *Excellent* naval gunnery school, 1830. (Collingwood Museum)

In 1729, the Admiralty decided to establish a purpose-built Royal Naval Academy in Portsmouth Dockyard. Entry to the Academy was by examination and on successful completion of training the 'King's Letter' boys were found a place in a ship. This method of entry resulted in a less random quality of officer being offered to a captain but did not displace the 'Captain's Servant' scheme and, up until 1794, a ship's captain was still authorised to have four servants for every 100 personnel in the ship's company. One reported example of this recruiting practice was that of Lieutenant James Cook who, as Captain of HMS *Endeavour,* registered his sons, James and Nathaniel, as his servants on board before departing for his first circumnavigation of the world in 1768. At the time, James was six and Nathaniel only five years old.

Unfortunately, from its origins the Academy suffered from a largely unjustified reputation that it was not a 'Gentleman's way of gaining a commission'. This slight was usually postulated by senior officers who considered that birthright was the main qualification for a naval officer notwithstanding that there were numerous examples of Academy graduates who went on to have very distinguished naval careers. The Academy became the Royal Naval College in 1806 and continued to operate as a source of officer cadets despite opposition at senior levels, including Lord St Vincent who described it as 'a sink of vice and abomination'. It was the introduction of HMS *Excellent* in 1830 as a harbour training ship, capable of sail and gunnery training, that eventually undermined the role of the College and its detractors used the financial argument to have it closed down in 1837.

Officers Training Ship

It was not until 1857, 20 years after the closure of the Naval College, that the Admiralty brought the recruitment of officers under full control. With the publication of Admiralty Circular Number 288, centralised training was introduced for all naval cadets on board HMS *Illustrious,* a two-decker ship of the line moored permanently in Haslar Creek off the entrance to Portsmouth Harbour. In this dedicated training ship, nominated boys undertook basic training in the duties of the Military Branch officer who was by now also referred to as a deck or executive officer in order to more explicitly indicate the nature of the duties involved. In 1859, owing to increasing cadet numbers, training was relocated to HMS *Britannia,* a three-decker, also at Portsmouth. However, as a local news report observed in November 1861,

> The *Britannia* is moored in the most unhealthy part of Portsmouth harbour and has been kept there in spite of some isolated cases of fever which have occurred, and which at the time raised

a great outcry from the friends of the youths on board. This obstinacy of officials in keeping the ship moored in a position that was, to say the least, of doubtful salubrity, has been probably justified by the fact of the average sickness on board the *Britannia* having been less than in most of the public or large private schools. As might have been foreseen, this perverseness is at length forced to give way, and the ship, instead of being moved to a more healthy position in the port, is now to be removed from the port altogether.

As reported, the public pressure eventually led to the Admiralty considering many alternative options for relocating the *Britannia* and, in 1862, the ship was moved for a brief sojourn at Portland before finally being transferred to Dartmouth in September 1863. Further expansion of cadet numbers caused HMS *Hindostan* to be moored alongside in 1864 to increase the accommodation space and, in 1869, *Britannia* was replaced by the larger HMS *Prince of Wales*, although the *Britannia* name was retained.

The relocation to Dartmouth of the *Britannia* did not prevent further epidemics on board and the health concerns associated with the use of wooden hulks for training and accommodation continued. These poor conditions were caused by the static and enclosed nature of the training hulk and aggravated by the location of the mooring sites, which, in addition to discharge of

Britannia and *Hindostan* in Dartmouth Creek, 1890. (Collingwood Museum)

HMS *Britannia* officer cadet class of 1887. (BRNC Museum)

sewage from the ship itself, were often adjacent to sewage outflows in badly polluted harbours. Another factor, which was certainly considered in the relocation of *Britannia*, was ease of access to the perceived temptations open to young men in the nineteenth-century port environment. This toxic combination became manifestly obvious when the sickness and mortality statistics were compared with those from the seagoing fleet, where the introduction of steam, iron hulls and electrical technology was leading to marked improvements in ventilation and hygiene facilities and rapidly reducing levels of sickness.

In 1875, a committee under the chairmanship of Rear Admiral E.B. Rice reported on the living conditions in *Britannia* and recommended that a college be built ashore. It was not until 1900 that a contract was finally let for the construction of a college and the hulks continued as a

Britannia Royal Naval College, 1981. (Collingwood Museum)

The hulk *Britannia* finally leaves Dartmouth, 1916. (Excellent Museum)

base for officer training until September 1905 when the Britannia Royal Naval College (BRNC) was fully opened for training. Many of the supporting facilities were progressively released for use before the completion of the main buildings in 1905, with the most notable of these being the hospital, which was opened in 1902, reflecting the importance of such a facility for the welfare of the trainees.

The opening of BRNC marked the first move of Royal Navy training ashore, something which was not completed until the closure of the hulk *Defiance* in 1954. *Britannia* continued to be used for ship's company accommodation and storage until 1916, when she was finally towed out of Dartmouth Creek in July for breaking and recovery of scrap metal needed for the war effort.

Selborne-Fisher Officer Training Scheme

Towards the end of the nineteenth century, following the surge in engineering and electrical technology hitting the Fleet, concerns were raised relating to the quantity and quality of officer manning. The skills emphasis had clearly moved away from the sailing of a warship into an era where knowledge of the new propulsion systems and weapons was of more relevance. Matters would become worse with the imminent arrival of the new WT systems. Notwithstanding the deployment of technology, one of the issues of major concern to the Admiralty was the apparent reluctance of the Victorian upper classes, which in the main formed the officer recruitment base, to become associated with engineering matters. When Admiral Fisher became Second Sea Lord in June 1902, he responded to these problems by announcing a reform of the officer corps in conjunction with the First Lord of the Admiralty, then the Earl of Selborne. Fisher's proposed reforms introduced a radical new approach to the structure and training of the Royal Navy's officer corps and it became known as either the Selborne-Fisher Scheme or the Common Entry Scheme.

The scheme proposed a single entry and training system for executive, engineer and marine officers with the idea of creating 'a body of young officers who at the moment of mobilisation for war would be equally available for all general duties of the Fleet and to consolidate into one harmonious whole the fighting officers of the Navy.' The implications being that all such officers would be considered as part of a Military Branch, appear in a single Military List and all be eligible for command. As such, all graduates of the scheme were to wear the same uniform with the executive curl[61] as part of their gold distinction lace and to have a specialist designation letter in brackets after their rank. At the time, for the executive officer, these specialist options were (N) for navigation, (T) for torpedo or (G) for gunnery. The engineer officers were to have the suffix (E) and the marine officers the suffix (M). Under Fisher's 'all of one company' concept each specialisation would receive sufficient common training such that on promotion to commander, both executive and engineer officers would have the option to revert to full time executive officer duties with the opportunity of gaining a ship command. Behind the inclusion of the marine officers in the scheme was the fact that Fisher was convinced that they were underemployed in ships for the majority of the time. As such, he considered that they could be used in times of shortage to carry out some non-seamanship related executive duties under the title of Lieutenants (M)[62] while they were on board ship.

The Selborne-Fisher scheme principles were not entirely original. Admiral Sir Astley Cooper-Key, as President of the newly formed Greenwich Naval College for Military, Engineer and Constructor Officers, had recommended equality of pay, rank and promotion for both Military and Engineer Branch officers as early as 1875. In addition, he had advocated that engineers should be re-categorised from the Civil to the Military Branch. This general theme had been subsequently resurrected by others but it was to take a man with Fisher's resolve to implement such controversial change.

The new scheme made no reference to the future of the existing Paymaster Branch. This was probably because Fisher espoused the view that the paymaster position was not appropriate for a commissioned officer as the duties were 'not arduous and that, except for a few days at the end of each month and quarter, an hour and a half's work a day will easily cover them.' This opinion was apparently based on the fact that Fisher felt most of the pay and victual accounting work was carried out by rating writers and stewards who were a 'very trustworthy and reliable class'. It is allegedly reported that Fisher also felt that commissioned officer responsibility for public funds could be undertaken by executive lieutenants who had been passed over for promotion and would carry the suffix (P). The corollary being that aides to flag officers should in fact be executive officers, not promoted paymaster officers.

The Fisher notion of giving command to a non-executive officer specialist was strongly resisted by supporters of the earlier Military Branch and grumbled at by the existing Engineer Branch because the opportunities were not to be retrospective to all engineers, only those who had been Selborne-Fisher trained. In 1903 in order to appease the existing engineers, Fisher changed the Engineer Branch titles, which had been functional up until that time, to align with the Selborne-Fisher Military Branch rank structure but with the prefix 'engineer', so that fleet engineers were given the rank of engineer commander. In 1905 when he became First Sea Lord, Fisher further proposed that the pre Selborne-Fisher Engineer Branch officers should also be granted Military Branch and Military List status, wear the executive curl and cease to wear the purple branch cloth between the rings of their gold distinction lace. However, this proposal was not accepted and it left the divisive situation of senior engineer officers appearing on the Civil List while their newly recruited subordinates were on the Military List.[63] Further pressure from the executive officer specialists eventually led to Fisher rescinding the idea of even Selborne-Fisher engineers being given the option for command on promotion to commander. A compromise was introduced whereby, at two points during time served as a lieutenant, the officer could opt to specialise as an engineer but to do so would mean forgoing all rights to revert to executive duties and to have command.

Royal Naval College Osborne

The lack of accommodation in the hulks at Dartmouth meant that the first Selborne-Fisher Scheme entry could not be accepted at BRNC in 1903. King Edward VII agreed that part of Queen Victoria's residence at Osborne House on the Isle of Wight could be converted to a training college, and in September 1903 the Royal Naval College Osborne was ready to accept the first batch of thirteen-year-old Selborne-Fisher officer cadets.

Entry to RNC Osborne was open to anyone by application to the Admiralty, with nomination to the College following successful completion of an interview, a qualifying examination and a medical. It was perhaps not a coincidence that the age of entry to Osborne coincided with that for public schools, thus presenting the College as a clear alternative education for the type of candidate being sought by the Royal Navy. There were also fees to be paid for this education but, unusually, the Navy did not accept any obligation to allow the student to continue studies regardless of performance. Although the selection process was rigorous, focusing on ability, character and social graces, the suitability interview was conducted by naval officers and any nomination had to be approved by the First Sea Lord. Arguably under these conditions, there was a danger that the 'suitable' type of recruit would not depart too much from the self perceived image of the military officer. Thus the anachronisms of the old nomination system were still likely to be prevalent unless suitably qualified and motivated engineers were included on the interview board and a progressive First Sea Lord was in post. With this variable in the system, the intention of increasing recruitment numbers by broadening access to the aspiring social classes, who would accept engineering as a career its own right, was also placed under threat.

Officer cadets spent a total of four years undergoing shore training with the first two years of common academic training provided by RNC Osborne and the final two years at BRNC Dartmouth. Throughout the whole period, academic study with a significant engineering bias was a key feature of the syllabus but so also was exposure to the naval ethos and the more practical elements of seamanship and naval living. This shore time was followed by three years at sea as a midshipman before taking a promotion board for acting sub lieutenant. Every officer had to then

Osborne cadet sail training in HMS *Racer*, 1905. (BRNC Museum)

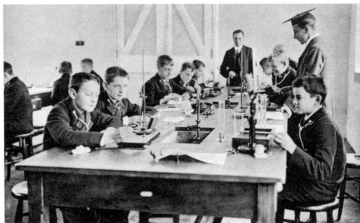

Above left: Osborne cadets recovering from a mumps epidemic, 1906. (BRNC Museum)
Above right: The laboratory of Osborne Royal Naval College, 1906. (Naval Museum Portsmouth)

take the engine room watchkeeping certificate before being given a first option to specialise as an engineer. Prospective engineers had to spend at least two years at sea as a sub lieutenant before returning to Greenwich College for academic refresher training and thence the Royal Naval Engineering College at Keyham.

Under the Selborne-Fisher scheme of training, the first engineer to appear in the military section of the Navy List was Lieutenant (E) Alexander Livingstone Penrose Mark-Wardlaw Royal Navy, who had a seniority of 30 June 1912 and was also a qualified submariner.

Wartime Changes

Following his recall out of retirement to return as First Sea Lord in October 1914, Fisher seized the moment and issued AFO814/14 on 24 December 1914 announcing that all officers in the existing Engineer Branch were to be considered part of the Military Branch, but only Selborne-Fisher graduates were to retain their eligibility for command. They were to wear the military officer uniform incorporating the executive curl and were also to retain their Engineer rank prefix and to continue to wear the purple branch cloth. The AFO stated that 'There was to be no change in their status as regards to the command of His Majesty's ships.' This announcement was followed by an Order in Council dated 7 January 1915 which instituted a new five-branch structure comprising Military, Medical, Accountant, Instructor and Artisan branches. This new structure was reflected in the April 1915 Navy List by showing the authorised number of Royal Navy officer posts broken down to reflect the five-branch structure with pre Selborne-Fisher engineers being included within the Military Branch list but still as a separate group. The supporting change to the Uniform Regulations was introduced in the July 1915 edition of the Navy List[64] where only the five-branch structure was shown and the engineers were allowed to wear the executive curl.

The next significant move towards Fisher's concept of a coherent officer corps came with the issue of AFO3224/18 dated 3 October 1918 and AFO3502/18 dated 1 November 1918. Under these directives the Accountant, Medical and Instructor branches were also given the military rank structure titles prefixed by paymaster, surgeon and instructor respectively. They were also permitted to wear the uniform and executive curl of the Military Branch differentiated only by coloured branch cloth stripes between their distinction lace. The reason given for this change was that the public persona of naval officers was being confused by the different uniforms and titles and that this had caused embarrassment during the war. This, coupled with the fact that the newly formed Royal Air Force had followed the Army lead in not differentiating specialist officers by title or uniform, meant that the time had come for the Royal Navy to recognise that its view on Military Branch status was essentially divisive. These changes were incorporated into the Uniform Regulations in January 1919 when the rules for officers indicated that only executive officers and Selborne-Fisher engineer officers were not required to wear coloured branch distinction cloth.

Unfortunately, while Selborne-Fisher succeeded in establishing a common officer training system, it did not explicitly lay to rest the issue of engineers being eligible for command. This debate continued to rage with ongoing and forceful representations at Admiralty Board level designed to protect the status of the executive officer. One memorandum put to the Board posed the question:

> Is it or is it not for the good government of H.M. Navy that officers of different Branches, who by the nature of their employment are non-executive officers and can have only limited responsibility, should be invested with the same power, should be called by the same titles and be dressed in the same uniform and thus become indistinguishable from the Executive Officers of the ship, who are responsible for the fighting efficiency, the general discipline and for the faithful discharge of their duties by every branch of officer and rating in the ship?

Perhaps more importantly, the Selborne-Fisher Scheme still did not resolve the problem of actually getting career officers involved in new technology and volunteering as engineers. Although many officers found the engineering training fulfilling, when faced with a choice between commanding a ship as an executive officer or not commanding a ship as an engineer officer, many found command more inviting. There is some evidence that this view was still being encouraged by parents whose ambitions for their offspring did not involve engineering. Another possible reason was that the selection process for 'suitable' individuals may have filtered out those without any preconceived notion of wanting a command. For these reasons, and that the scope of the training was found to be too broad for many individuals, the Selborne-Fisher Scheme was eventually acknowledged to be unworkable. In 1921, the decision was made that engineer officers and executive officers were not interchangeable and the scheme was finally stopped. In 1922, engineer officer training was revised such that officers joined RNEC Keyham as midshipmen for professional training that extended for three years and took them up to the rank of lieutenant. The first batch of specialist engineer officers qualified in 1925 and engineering training for executive officers was phased out.

HMS *Britannia*
officer cadet
class of 1901.
(BRNC
Museum)

The status quo became fully established in November 1925 when the Admiralty issued an Order in Council, 139/CW dated 16 December 1925, which abolished the branches established in 1915 and put in place a total of twelve new branch categories of officer. Seven of these categories were branches applicable to commissioned officers and they were the Executive,[65] Engineer, Medical, Dental, Accountant and Instructor Branches, plus the Chaplains. The remaining five branches were designated for warrant officers only and comprised the Schoolmaster, Wardmaster, Shipwright, Ordnance and Electrician specialisations.

The Order made no reference to a Military Branch and consolidated the establishment of a separate Engineer Branch which was to comprise all engineer specialists, both pre and post Selborne-Fisher. The branch was to be equivalent in all general respects to the Executive Branch except for being eligible for command. Once again this constraint was not explicit, possibly because it would be applied through the appointing system by the categorisation of command appointments as for executive officers only. Although the Order inferred the final demise of the Military and Civil Lists, as Admiral (E) Louis Le Baily recalls, soon after the first engineer officer joined the Board of Admiralty as Fourth Sea Lord, a notice by the Navy Club inviting the Admiralty Board to dinner was addressed to 'The First and Second Sea Lords, The Controller of the Navy and other Civilian Members of the Board' which clearly implied the perceived status of the Fourth Sea Lord, a specialist engineer by profession, as being civilian in nature.

The message sent out by the 1925 Order remained unequivocal in that only the Executive Branch could aspire to command and, as if to reinforce that point, at the same time there was also an Admiralty Fleet Order instructing officers, other than executive officers, to wear coloured branch cloth between their stripes, with all engineer officers now being required to wear purple.

Seaman Ratings

When the numbers of royal ships were few, most of the seamen to be found on board ships brought into the King's service were manned under the Fyrd. The majority would have been merchant seaman, probably being provided along with the ship in which, during peacetime, they made their living. There would also be a few landsmen to bring up the numbers and to fill the artisan trade positions on board, such as carpenters, coopers and blacksmiths. These artisans would all be retained on board, even when alongside, for as long as their ship was in the King's service. This may have been where the term ship's commission first arose with the crew being paid off at the end of their obligated duty period.

As the number of Royal Ships started to increase during the fifteenth century, the Royal Navy and the merchant fleet found itself increasingly in competition for manpower during peacetime as well as in wartime. The Fyrd was no longer a viable way to man a standing Royal Fleet and this led to the introduction of impressment which was a statutory right for the state to forcibly recruit men into either military or naval service. The reason given for enacting the first impressment statute was initially as a curb on vagrancy and the consequence was that anyone found guilty of such a charge in a port area was invariably drafted into naval service. The interpretation appears to have loosened over time owing to the enthusiasm and desperate need of the press gangs, which led them to pressing any suitable man found in the port areas at the wrong time.

Another ploy used by the press gangs was to intercept merchant ships returning from a trading voyage, press a number of the crew into service and use a naval crew, who could be trusted to return to their own ship, to complete the final leg of the merchantman's voyage back into home port. Given the financial penalties of losing a good crew, this tactic was resisted by the larger trading companies such as the British East India Company. Many of these companies armed their vessels against privateers and pirates but they are also on record as having fired on naval cutters carrying press gangs and attempting to board them on their return to England.

A number of attempts were made to set up a register of seamen but these were unsuccessful and in 1793 the Impress Service was established ashore as a uniformed recruiting arm for the Royal Navy. The activities of the Impress Service caused a great deal of resentment amongst civic leaders in the port areas around the country and riots were not unknown when the press gangs were on the prowl. Not all pressed men felt strongly aggrieved about being kidnapped into the King's service, particularly the skilled seamen who frequently regarded a period spent in the Royal Navy as an opportunity to be rated as a petty officer and gain a small pension. Their only real problem was that the time spent in naval employment was unpredictable and the sudden end of a war quickly meant the termination of their employment and discharge ashore. Unlike officers, who were similarly discharged ashore but on half pay, the seaman was often owed back pay and had to find alternative work quickly. Once pressed into service, the seaman would be interviewed on board his new ship and rated according to his abilities and experience. This rating could be at Ordinary Rate, Able Rate or, in some circumstances, Petty Officer Rate, although the latter was more likely to be given to a man once he had proven his loyalty and capability to his ship's officers.

Active impressment ceased after about 1815 but the statutory right remained on the books if needed. In 1835, any man who had completed five years naval service was relieved of any threat of being pressed. Finally in 1853, the Navy brought in continuous service engagements which went a long way to removing the uncertainty of the old system of service and guaranteed the man a pension after a fixed number of years before the mast. Although these obligations were occasionally allowed to lapse, notably in 1857 after the Crimean War, they did form the basis of the conditions of continuous service as they have evolved until the present day. Although improvements were gradually being introduced, recruiting was still very much on a ship-by-ship basis as the poster for HMS *Algiers* from 1860 (opposite) shows.[66]

Most seamen learnt their trade working on board ships under the guidance of their officers and petty officers. The Royal Navy had also introduced a number of brigantines, two-masted square rigged ships, which sailed the coast giving young men the opportunity to taste life at sea. These served as mobile recruiting units and provided a basis for sail training at sea. There was little formal shore training until 1830 when HMS *Excellent* was commissioned as a training school in Portsmouth. Although the main purpose of *Excellent* was gunnery training, the ship was also used to train seamanship skills and give practice in the working of a ship's sails and rigging. In 1854, the hulk *Illustrious* was set up as a harbour training ship for boys at Portsmouth. This was followed the next year by the *Implacable*, which was moored in Devonport. It was the use of the *Illustrious* for boys' training that eventually led to the concept being extended to officer cadets in 1857.

An HMS *Algiers* recruiting notice, 1860.
(Collingwood Magazine Archives)

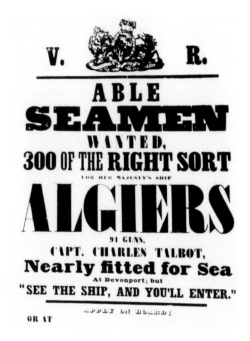

By 1860, the days of the sailing ship were numbered and sail training was no longer a priority, although many of the associated seamanship skills would continue to remain a requirement until the present day. Apart from seamanship duties, after 1860 the seaman rating was left with one expanding area in which he could specialise: that was in gunnery.

The Gunnery Branch

Some of the earliest ships built for the specific purpose of naval warfare – as opposed to armed merchant vessels co-opted into the King's service – were produced around 1420. In addition to small arms, these ships carried early forms of cannon, sited on the castle decks at either end of the ship. They were unreliable at best and, at worst, prone to explode on firing. As early as 1344, there had been a Clerk of the King's Ships responsible for assembling a naval force as required in defence of the realm but no statutory figurehead with responsibility for naval ordnance. There did exist an Ordnance Office, which supported primarily the Army with the Royal Ships receiving less attention due to their emerging status. It was not until 1545 that the post of Master of Naval Ordnance with responsibility for naval guns came into being as a member of the Council for Maritime Causes, the forerunner to the Navy Board and, subsequently, the Board of Admiralty in 1832.

Since 1420, gun technology had been mainly focused on improving reliability and safety in operation. Although the technology for casting barrels had been improved in the early 1500s, the principle of the cannon had changed little up until the eighteenth century, other than increases in size of the shot weight. One innovation that had been introduced around 1505 was the use of gun ports which allowed cannon to be deployed down each side of the ship and under cover, rather than exposed on castles as in the case of earlier warship designs. Eventually this invention led to the use of synchronised firing, known later as broadsides, when engaging the enemy. This tactic started to emphasise the need for captains to manoeuvre their ships into the best firing position, having regard to wind direction, sun, lee shores and any other factors which might frustrate the enemy's attempts to fight back or prejudice the chances of their own success, particularly by blinding themselves with their own gun smoke.

Even the early naval guns and small arms required some form of maintenance to keep them in a safe working condition. It was this requirement which, arguably, led to the founding of the first specialist technical branch in the Royal Navy. This came with the introduction of a gunnery specialist in 1731, whereby the gunner was given responsibility for a department comprising the armourer and the gunsmith onboard the King's warships under a Board of Ordnance Warrant. The Gunner was probably a petty officer rate until 1778 when he was given warrant rank and appeared on the Military List.

Board of Ordnance Warrant c. 1731
The Armourer and Gunsmith are straitly charged to observe the Gunners orders, and their special duty is to attend the small arms keeping them clean within as well as without, by frequent oyling them; but not to take them too often to pieces which destroys the locks, screws and other parts.

While the 1731 edict gave guidance as to onboard responsibility for naval guns, it still did not address the training of the Fleet in gunnery. In fact, how much time was devoted to the conduct of gunnery drills varied greatly from ship to ship depending on how much of his private purse the captain wished to invest in live firings in order to increase the opportunities for gaining prize money.

It was only after some poor gunnery performances during the Napoleonic Wars and, particularly, some disastrous engagements with the US Navy during the war of 1812 that the Navy Board decided to establish the shore-based Gunnery Training School for the Fleet on board HMS *Excellent* in 1830. The direction given to the Commander-in-Chief Portsmouth by the Navy Board for the establishment of a gunnery school stated:

> Their Lordships having had under their consideration the propriety and expediency of establishing a permanent corps of seamen to act as Captains of Guns, as well as a Depot for the instruction of the officers and seamen of His Majesty's Navy in the theory and practice of Naval Gunnery, at which a uniform system shall be observed and communicated throughout the Navy, have directed, with a view to such an establishment, that a proportion of intelligent and active seamen shall be engaged for five or seven years, renewable at their expiration, with an increase of pay attached to each consecutive re-engagement, from which the important situation of Master Gunner, Gunners' mates and Yeoman of the Powder room shall hereafter be selected to instruct the officers and seamen aboard such ships as they may be appointed to in the various duties at the guns, in consideration of which they will be allowed two shillings a day per month, in addition to any other rating they may be deemed qualified to fill, and will be advanced according to merit and the degree of attention paid to their duty, which if zealously performed will entitle them to the important situations before mentioned, as well as that of Boatswain.

The implications were far reaching and, possibly for the first time, the Navy Board was starting to address issues such as terms of engagement, advancement opportunity, qualification pay, pay increases based on time served and merit as well as the concept of standard Fleet operating procedures. Another significant fact was that the master gunner retained eligibility to be rated boatswain, a warranted rank on the Military List, which meant that he could also aspire to command.

Before commissioning, *Excellent* was fitted with guns of every type used by the Fleet. It was then used to conduct live firing drills for prospective gun crews, albeit while anchored in Portsmouth Harbour and firing over the mudflats. Apart from firings, the school provided instruction in gunnery theory and the maintenance requirements of individual weapons including

> … the names of the different parts of a gun and carriage, the dispart in terms of lineal magnitude and in degrees how taken, what constitutes point blank and what line of metal range, windage and the errors and the loss of force attending it, the importance of preserving shot from rust, the theory of the most material effects of different charges of powder applied to practice with a single shot, also with a plurality of balls, showing how these affect accuracy, penetration and splinters, to judge the condition of gunpowder by inspection, to ascertain its quality by the ordinary tests and trials, as well as by actual proof.

It is worth summarising the position of the gunnery specialist in the first half of the nineteenth century in terms of technology. The science of accurate gunnery had been around for some time but few naval officers had serious interest in applying it to their operations, mainly because the most popular tactic was to engage the enemy closely and exchange broadsides until one or the other of the combatants struck their colours. This modus operandi did not necessitate much in the way of ballistics knowledge but it did encourage more rapid rates of fire in a safe manner which would avoid killing the ship's own crew or at worst sinking one's own ship. The training at *Excellent* started formally to spread the art of gunnery and, in general, develop the Fleet's capability. Even so there were still some senior officers ashore and afloat who doubted whether the cost of

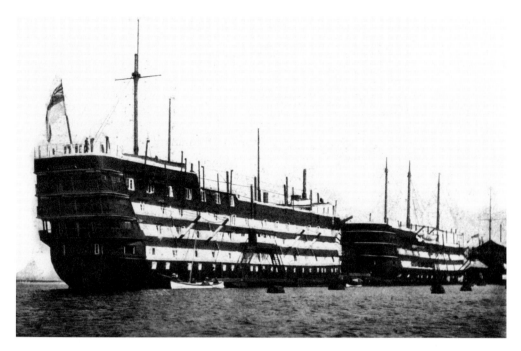

The Third HMS *Excellent* naval gunnery school *c.*1860–1891. (Collingwood Museum)

the training was worth it, no doubt forgetting the lessons learnt from the 1812 war against the US. With the knowledge of ballistics now required and the drip of technological improvements to the gun in the form of new firing mechanisms, breech loading, powered training and elevating systems and more sophisticated ammunition design, the gunnery specialist was required to be more technically capable.

The Engineer Branch

Engineers first appeared in Royal Navy ships as maintainers of fairly complex equipment in 1819 when the first steam-propelled ship, *Comet*, came into service as a tug. *Comet* was followed by HMS *Lightning* in 1823 and, thereafter, steam-propelled ships started to become a common sight. Engineers were not highly regarded at the time and had a status equivalent to the traditional trades, such as carpenters, coopers and shipwrights. These tradesmen were collectively categorised as 'artisans' because of the additional trade skills they required to carry out their duties. As far as rank was concerned, the engineer was of petty officer status but, given his fundamental importance to the ship, this low rank was of concern to the ship's deck officers. The reason for this concern was that it was felt that the engineer's loyalty to the ship might be found wanting unless rewarded with more public recognition of his role. In a bid to bring some recognition to the Engineer Branch, the best qualified engineer on board was usually designated as the First Engineer and given the authority to allow him to control an engineer department.

In the early 1830s, it was clear that the Royal Navy was committed to steam propulsion and a fully equipped steam factory was opened at the Woolwich Dockyard. This led to similar facilities being opened in Chatham, Portsmouth, Devonport and Sheerness. The Royal Dockyards started to train their own engineering apprentices and it was largely from this source that the Royal Navy recruited its engineers.

In July 1837, an Order-in-Council formally created the Engineer Branch Afloat and engineers were given warrant officer status on the Civil List and accorded the ranks of First Class, Second

Class or Third Class Engineer. They were messed in the gunroom with other officers below the rank of lieutenant. The Order also introduced the first Royal Navy Engineer Apprentice scheme, which involved five years of training, including three at sea and two in the Royal Dockyards.

In 1843, First Class Engineers were given commissioned rank and in 1847 another Order-in-Council announced a new Engineer Branch structure comprising the ranks of Inspector of Machinery Afloat, Chief Engineer and Assistant Engineer. This edict gave engineers an equivalent rank to that of other Civil List officers on board, such as surgeons and pursers. As Civil List officers, they were still not considered to be wardroom status. In 1863, the engineers' uniform carried for the first time a cloth purple branch stripe alongside the gold distinction lace stripe(s) designating their rank. At around the same time, other specialist Civil List officers were also given branch designating colours but neither they nor the engineers were allowed to wear the gold lace executive curl or eight button double breasted uniform, both of which were reserved for Military List officers.

The final nail in the coffin of the sailing ship came when it was officially announced in 1859 that 'Sailing ships were unfit for active service' and by then the Keyham Steam Factory at Devonport was in full production. On 1 July 1880 the Devonport Dockyard Training School for Engineers, shortly afterwards to be renamed the Royal Naval Engineering College, Keyham, was opened and it started training the Royal Navy's engineer officers on 4 July 1880. When the Selborne-Fisher Scheme came into being in 1903, newly recruited engineer officers at last started to receive Military List status and became part of the common officer training programme. The introduction of this programme led to a break in the training pipeline which caused Keyham College to cease training in 1910 and re-open three years later to take the first of the Selborne-Fisher trainees. When Selborne-Fisher was terminated in 1921, it was replaced by the Long Engineering Course, a four-year course, which ran until 1951. For much of this time the engineer officer had responsibility for the electrical equipment inside his department's machinery spaces, except for the generator end of any diesel-driven or steam-driven prime mover. For the support of his equipment, the officer had only the limited electrical skills of the Engine Room Artificers (ERA).

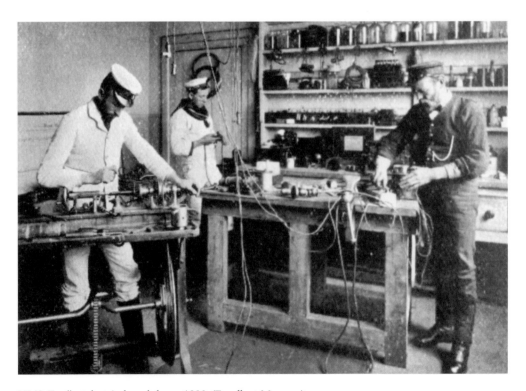

HMS *Excellent* electrical workshop *c.*1890. (Excellent Museum)

ERAs had been introduced as the Royal Navy's skilled engineering tradesmen in 1868 and they worked for the ship's Engineer Officer as part of the Engineer Department. Along with the steady progress in steam technology came the first use of electrical power generation machinery in the guise of the dynamos required to power the new searchlights. In order to cope with new area of technical responsibility, in 1881 the Engineer Branch started electrical training at the recently opened Torpedo School in HMS *Vernon* with engineer officers being given one week and ERAs being given four weeks of formal instruction.

Initially, an ERA had been trained as an apprentice in the Royal Dockyards but in 1902 Fisher introduced a scheme for training Boy Artificers and established training schools in each of the three port areas, Chatham, Devonport and Portsmouth with the latter being based in a collection of hulks designated as HMS *Fisgard*. A letter dated 29 December 1902 from the Commander-in-Chief of the Nore Command directing the establishment of the Boy Artificer training facility at Chatham began:

> My Lords Commissioners of the Admiralty having decided to introduce a system of entering Boy Artificers with the object of training them in the reserve for the rating of Engine Room Artificer, R.N., I am commanded to forward the enclosed copy of the draft regulations which have been prepared in order to give effect to their Lordships decision. My Lords propose, in the first instance, to introduce the new system in the Reserve, at Chatham, and to ensure an early start in the training of these boys, arrangements have been made with the Supt. of Greenwich Hospital School for the selection of 20 boys of the age of 15-16, with the view of sending them to the Depot at Chatham on the 19th of January next. Copies of this letter and enclosure addressed to the Supt. of the School are sent herewith for your information as a guide.

By the early 1920s, the training of all artificers became concentrated on *Fisgard* at Portsmouth before it was moved to Chatham where the trainees were accommodated ashore until 1939. The onset of war meant a rapid increase in the required number of artificers and this, and the threat of bombing in the Chatham area, led to two purpose-built training schools being established at Rosyth in Scotland, HMS *Caledonia*, and Torpoint in Cornwall, which retained the name *Fisgard*.

Manning for the Electrical Era

By the late nineteenth century the use of electricity had been firmly established in the Royal Navy through the introduction of electric lighting, searchlights and firing systems for both torpedoes and guns. With this expansion of electrical technology came the need for a new type of specialist and the problem of establishing a shipboard organisation to support the Fleet. The choice was either support by the end user, such as the Gunnery Branch, or support by a specialist electrical maintainer under the technical control of the Engineer Branch.

The case for the user maintainer was that the new technology was part of the military capability of the ship and therefore should be the responsibility of the Military Branch. Following the precedent set when it took over ordnance maintenance tasks, the Gunnery Branch considered that it should take on the responsibility for the new electrical systems, absorb the new technology and extend its field of expertise.

The case for the specialist maintainer was that the Engineer Branch had been maintaining a variety of ship equipment since 1819 and had an engineer officer training school in place at Keyham, in Devonport, which was working closely with the Royal Dockyards in the electrical field. The Branch also had an underpinning technical rating structure of ERAs and stokers. Thus it was that the Admiralty Board first had to decide its philosophy for electrical technology support and this decision was to be constantly reviewed from that time onward.

The Creation of the Torpedo Branch

Eventually the Board decided that the Gunnery Branch, which by now had its training school well established on board HMS *Excellent* at Portsmouth, was to take responsibility for the new electrical technology and it was tasked with providing officers and ratings for duties associated with the operation and maintenance of torpedo and mine warfare systems. In recognition of the new underwater warfare duties, a new warrant officer rank of Torpedo Gunner was introduced into the Gunnery Branch. The ratings used to support the Torpedo Gunner were drawn from existing seamen gunners and these ratings were given the designation 'Seaman Gunner Torpedomen'. With the new categorisation came the formal introduction of what was known as the non-substantive rate, in addition to the substantive rate, a decision which was to provide the basis for many a mess debate in the future.

Essentially, the substantive rate was a military rate which identified the bearer as a superior officer under the terms of the Naval Discipline Act (NDA)[67] and gave the bearer the right to give orders to any junior rate or officer. It was an authority initially vested at the leading rate, with increasing authority upwards through senior rate, warrant rank and commissioned rank. However, the non-substantive rate was a professional specialist qualification with the highest level being generally known as a First Class Rate. Thus, for arguably the first time in the Royal Navy, the new torpedo structure created a situation whereby a chief petty officer could hold a specialist qualification of Torpedoman 1st Class, which was a lesser qualification than that of Leading Torpedo Operator (LTO), which could be held by a leading seaman. In situations where technical competence was demanded, this could clearly lead to some confusion in the chain of command.

On 16 July 1872, then Commander J.A. Fisher was appointed Chief Torpedo Instructor at *Excellent* with responsibility for all training in torpedoes, mines and demolitions. In 1873, the electrical firing circuit technology being introduced into naval guns was also added to the training remit. From an electrical point of view, the training was all about the use of battery power to detonate ship-borne torpedoes, fire guns or initiate sea mines being used to protect harbours and anchorages.[68] Seaman Gunners were given this torpedo training in conjunction with their gunnery training but both aspects continued to be subsidiary to seamanship training which was still considered to be the primary professional skill requirement.

The links with the Gunnery Branch were not destined to last. In 1873, the old hulk *Vernon* was made a tender to *Excellent* and was designated as the Royal Navy's Torpedo School. The first torpedo long course was established shortly afterwards in *Vernon* and the hulk *Ariadne* was moored alongside to provide accommodation facilities. In 1876, this led to separation from the Gunnery Branch and the creation of a new autonomous Torpedo Branch. The Torpedo Branch became a part of the Military List and Torpedo Officers could achieve military command. The branch was given responsibility for all matters connected with underwater warfare, including the development of underwater weapons and their tactical and strategic uses. The importance of this embryo area of warfare was finally reflected in the fact that in 1876 HMS *Vernon* was commissioned as an independent command from *Excellent*. By 1879, *Vernon* was made up of the hulks *Vernon*, *Ariadne* and *Actaeon*, with the *Vernon* hulk being replaced by the *Donegal*, a more spacious ship, in 1886.

When the Torpedo Branch became a separate entity from the Gunnery Branch, the Torpedo Officer was given a ship department in his own right and he was assisted by a Torpedo Gunner, or 'Gunner T', which was a warrant rank. In the 1880s, a typical torpedo department comprised the Torpedo Instructor, one or more LTOs plus a number of Seaman Gunner Torpedomen. A Torpedo Instructor had to hold the substantive rate of either chief petty officer or petty officer. The LTO was a non-substantive rate qualification given to those who showed an aptitude for electrics which enabled the holder to carry out practical maintenance and repair work on general electrical systems as well as torpedoes. Although the qualification carried the title 'leading' in fact it could have been held by any of the substantive rates of able seaman, leading seaman or petty officer.

HMS *Vernon* Portsmouth torpedo school, 1910. (Collingwood Museum)

Torpedomen on HMS *Vernon* c.1898. (Collingwood Museum)

As late as the 1877 Uniform Regulations, the Seaman Gunner Torpedoman was not formally given insignia which showed his training and skills in torpedo matters. It was only in the 1879 Uniform Regulations that the skill badges were revised to show a torpedo but even then the badge carried both the gun and the torpedo insignia, with the gun overlaid on top of the torpedo. It was not until a man became professionally qualified as an LTO that the badge changed to show the torpedo overlaid on top of the gun. It was in the 1879 Uniform Regulations that the term LTO was first believed to have been used. LTO was a term to be remembered as for the next 66 years the rate was to play a considerable part in the electrical maintenance activities of the Fleet. The next professional step was to Torpedo Instructor and even at this level, the gun continued to appear in the background of the non substantive badge.

In 1879, the Torpedo Artificer was introduced as a skilled rate of similar standing to the ERA. The Torpedo Artificer was responsible for the maintenance and preparation of the torpedoes but not the torpedo launchers which remained the responsibility of the Armourer in the Gunnery Branch. The rate was not based on any specialist electrical expertise and, in 1894, all torpedo artificers were recategorised as Chief Armourers and the Torpedo Artificer rate was withdrawn. The Armourer retained his combined gun and torpedo duties until the introduction of the Ordnance Artificer (OA) on 1 March 1919. An Order in Council, CW2032/1920 dated 22 January 1920, instituted the Ordnance Branch concurrently and along the lines of the Electrician Branch with artificer promotion opportunities up through Warrant Ordnance Officer to, eventually in 1923, Ordnance Lieutenant Commander. The initial source of OAs came from the armourers and engineer artificers who had been involved with ordnance support prior to 1919.

In 1881, Admiralty approval was given for the construction of an 'Electric Light Shop' for the purpose of electrical training. By now the Torpedo Branch was firmly established and because electricity had been chiefly associated with torpedo warfare including searchlights and mines, it was decided to build the facility in *Vernon* and give the Torpedo School the additional responsibility of providing electrical training for the Royal Navy. The training facility was established on board the hulk *Ariadne*, part of the *Vernon* training group, in response to Admiralty Circular 91G dated 4 June 1881:

> In each vessel fitted with electric light at least one commissioned officer of the Military Branch [besides the gunner and torpedo officers] and two petty officers are to be instructed in their use and application. At least one Engineer Officer and one ERA besides those who have been instructed in the Naval Torpedo School are to be thoroughly acquainted with the management and use of the generating machines.

The 'Electric Light' referred to in the directive was the searchlight, not internal ship lighting which was still in the early stages of development. The 'Generating Machines' were the dynamos used to power the searchlights. All other electrics at this time were the battery powered, or low power, electrics used in the firing circuits for guns and the detonation of mines and torpedoes. It was as a result of this directive that, in 1881, the Engineer Branch started to receive training in electricity generating machines at *Vernon*.

In 1882, all Torpedo Gunners and Torpedo Instructors started to receive four weeks electrical instruction and this was extended to LTOs the following year. In 1883, a Torpedo Training Committee recommended the establishment of additional Torpedo Training Schools at the other two home port areas of Devonport and Chatham.

In 1884, *Defiance*, one of the last 'wooden walls', was laid up at Devonport and commissioned as the Devonport Torpedo School. The ship was fitted out with boilers, a dynamo and workshop facilities and used for torpedo and electrical training of Devonport ratings. In 1892, the hulks *Perseus* and *Flamingo* were moored alongside *Defiance* to provide supplementary electrical power and steam as well as additional accommodation. The hulks that comprised *Defiance* evolved over time with more modern ships being introduced when there was a need to update power generation capacity and steam requirements for further accommodation and to add new training facilities such as those for WT, which by the early twentieth century was advancing very rapidly.

HMS *Defiance* Devonport Torpedo School, 1914. (Plymouth Naval Museum)

HMS *Actaeon* (ex-*Ariadne*) Nore Command Torpedo School, 1905. (Collingwood Museum)

A similar school was eventually set up for the Nore Command at Sheerness in 1905 when the *Ariadne* was released from the *Vernon* training complex at Portsmouth, moved to Sheerness and renamed HMS *Actaeon*. The Sheerness Torpedo School was later moved to Chatham.

By 1897, the Admiralty had decided that electrical installations had become important enough to warrant an electrical section to be added to the complement in a ship's company. To meet this requirement, Seaman Torpedomen started to be given general electrical instruction to support the

A ship's electrical section at work
c.1900. (Collingwood Museum)

LTO at sea and smaller ships, such as destroyers, were allocated a billet for a Torpedo Instructor. Around 1907 the term 'instructor' was dropped and the Torpedo Instructor became the Torpedo Gunner's Mate. Paradoxically, as the Torpedo Branch developed further, the junior ratings, who were originally rated as Seaman Gunner Torpedomen, lost the term 'gunner' from their title to make them Seaman Torpedomen with no mention of 'gunnery'.

Although they had received some electrical training, Torpedo Department officers and ratings were still seamen first and torpedomen second. Overall their training reflected a need for them to understand the working and operation of their equipment rather than provide electrical engineering expertise. Thus they were not trained in the engineering craft skills that were rapidly becoming a key requirement for supporting other electrical systems onboard ship and this problem needed to be addressed.

HQ Manning for Electrical Systems

At the turn of the twentieth century, the Board of Admiralty set up a committee to study the potential scope for using electricity in the Fleet with a view to establishing a headquarters organisation to properly support the implementation of electrical systems in ships. The study recommended the formation of an Electrical Engineering Department as a department within the Directorate of Naval Construction (DNC) and in 1903 this led to the appointment of an electrical engineer, Charles Wordingham, to Admiralty Headquarters.[69]

This subordinate role within the DNC did not reflect the organisation in the Royal Dockyards where the electrical department formed a separate department to that for naval construction and, by now, it was designing electrical installations for a wide range of uses in ships. However, things changed in 1918 when the decision was made to recognise the importance of the technology and form a separate Directorate of Electrical Engineering (DEE) at Admiralty Headquarters.

The Director DEE became the Electrical Engineering Advisor to the Board of Admiralty and he was made responsible for the design and production of all electrical machinery except those equipments that used electro-magnetic waves, which at the time comprised only WT systems. The Director's responsibilities also included the provision of assistance with the production and overseeing of electrical materiel being dealt with by other departments. Another key responsibility was for the installation of all electrical machinery and apparatus. This was a highly significant change as it effectively brought some form of electrical standards organisation into existence with the engineering departments in the various dockyards now having a coordinating Design Authority based at Headquarters.

The Electrician

By the late 1890s it had become apparent that more in-depth electro-mechanical skills were needed for the on-board maintenance of power generation and distribution systems and the many gunnery and miscellaneous electrical equipments being fitted into ships. The result was that Admiralty Circular N8258/1901 dated November 1901 announced the establishment of the Electrician Rate within the Artisan Branch.

The conditions of service for the Electrician were the same as those for the ERA in the Engineer Branch. After entering as an Electrician 4th class on a daily rate of pay of five shillings and sixpence, the man could expect to be drafted to a ship with the substantive rate of chief petty officer and reach the non substantive rate of Chief Electrician 1st Class within fourteen years; after this time his pay would have risen to seven shillings and sixpence a day. The Admiralty Order authorised the recruitment of 100 Electricians, evenly split across the three main naval port areas. Chatham and Portsmouth ratings were to be trained in *Vernon* and Devonport ratings in *Defiance*. As the required electrical engineering skill base did not exist in the Royal Navy at the time, most of the first recruits were poached from the civilian industrial field. Usually they were ex-apprentices, aged between 19 and 28, who already had the necessary skill of hand and craft ability[70] to allow them to undertake a six-month course on naval equipment on joining.

The arrival of the Electrician was not widely welcomed by many in the Fleet and, like his ERA colleagues in an earlier era, he was barely tolerated by a Service in which the smartness of ships and dress were a priority and anything resembling an overall was still frowned upon. Nevertheless,

Recruiting electricians in the West Country, 1901. (Collingwood Museum)

it is reported that the early recruits were 'so enthused with their trade that they suffered the initial abuse with good natured tolerance' and they gradually built a reputation for efficiency in a technology which was having a rapidly growing impact on day-to-day service life.

Initially, the career path offered to the Electrician was limited to Chief Electrician, but Admiralty Order in Council 12553/1910 dated 19 October 1910 introduced an Electrician Branch for warrant officers only as a promotion route for Electricians. The Commissioned Warrant Electrician and Warrant Electrician ranks were instituted and the first Warrant Electrician appeared in the Navy List in 1912. The total number of promotions was limited to 24, of which no more than 50 per cent could be Commissioned Warrant Electricians. Promotion from Warrant to Commissioned Warrant Electrician rank required at least ten years of service in the warrant rank and therefore was not a frequent occurrence.[71] By the beginning of the First World War, only a small number of promotions had been made to Warrant Electrician.

Specialist Branch colours had already been in force for warrant officers in other branches before 1912, but it was not until January 1919 that the Uniform Regulations required the Warrant Electrician to wear a dark green cloth ring under and adjacent to his thin gold ring. Commissioned Warrant Electricians wore the same green cloth stripe but under a thick gold ring, the same as a sub lieutenant. It would seem reasonable to assume that this introduction of the Branch colour could well have given rise to the first collective use of the term 'Greenies' for the ship's electrical section.

An Order in Council CW2032/1920 dated 22 January 1920 abolished the rank of Commissioned Warrant Electrician and replaced it with 'Commissioned Electrician'. In addition, the order further extended the promotion available to the Electrician by introducing the commissioned rank of 'Electrical Lieutenant' for selected commissioned electricians. A further order, CW4156/23 dated 16 April 1923, introduced the opportunity for promotion to 'Electrical Llieutenant Commander' after eight years' service as an Electrical Lieutenant. This higher rank was limited in numbers to 8 per cent of the number of Electrical Lieutenants. The new ranks were collectively referred to as 'Commissioned from Warrant ranks' and were still viewed as part of the warrant officer only Electrician Branch formed in 1910 and endorsed in 1925. This career structure remained in place for specialist electricians until the formation of the Electrical Branch in 1946.

The Electrical Artificer

Following an Order in Council N14305/1913 dated 24 June 1913, electricians below the rank of Warrant Electrician were re-categorised as Electrical Artificers (EA), on a par with the ERA, and took over the ERA's electrical duties.[72] The EA had no seamanship responsibilities and was directly responsible to the Torpedo Officer for his technical duties which covered the maintenance and operation of the ship's electrical systems. In 1914, the EA started to receive training in the Whitehead Torpedo and, in addition to the electrics, he took over the mechanical aspects of the weapon which had previously been the province of the Armourer and the ERA. The promotion route for the EA remained the same as his predecessor which was up through Warrant Electrician.

Prior to 1920, the EA had been recruited as an adult as there was no equivalent to the Boy Artificer entry in the Engineer Branch which had been in existence since 1901. An Order in Council, N9921/1920 dated 17 May 1920, introduced the 'Boy Electrical Artificer' scheme whereby a proportion of EAs should be recruited as boys. This scheme subsequently evolved into the Artificer Apprentice scheme used for all Royal Navy artificer technical training.[73] The scheme accepted school boys between the ages of 15 and 16 and put them through a common craft training programme before then allowing them to specialise as an engine room, ordnance or electrical artificer. In due course, it was this source of EAs that provided the basis for a ship's electrical section with support from electrically trained torpedomen. EAs continued to report to the Torpedo Officer in ships until the end of the Second World War at which time they came to form the core element of the new Electrical Branch structure.

The increased development of the electrical gyro with its many applications during the 1920s meant that the skills demanded of the EA were increasing to a degree where it was considered necessary to introduce the first form of sub-specialisation within the EA category. Accordingly, a number of EAs were selected for a special course in gyro compasses and navigational systems at the Admiralty Compass Observatory in Slough. These artificers were then drafted to billets with responsibility for gyro maintenance aboard larger warships such as battleships, carriers, cruisers and depot ships.

The Wireman

Soon after the start of the First World War, the Royal Navy found that it had not got the resources or skills needed for repairing electrical wiring damage. This led to the Electrical Wireman rate to be introduced under Order in Council N48049/1915 dated 10 November 1915. The electrical wireman was brought in for Hostilities Only and the personnel were used to support general electrical repairs. They were drafted into leading torpedoman and seaman torpedoman billets but borne for technical duties only and not obligated to carry out any seaman duties normally associated with such billets.

Support for Wireless Telegraphy

Following the success of Marconi's WT demonstrations to the Admiralty, the fitting of WT equipment into the Fleet proceeded at a great pace. By 1901, *Vernon* had become the focus for wireless communication development and this led to the foundation of the Wireless Telegraphy Experimental Establishment (WTEE) in 1905. The first *Handbook of Wireless Telegraphy* was published in 1901 and the torpedo schools in Portsmouth and Devonport both started regular WT training courses with 150 officers and ratings, drawn from the Torpedo and Signals Branches, being trained in the first year. As the operation and maintenance of WT equipment was taught at *Vernon*, the responsibility for WT onboard ship fell to Torpedo Officers. On 16 March 1906, Admiralty letter ALN/2683 announced the formation of the Wireless Telegraphy Branch as a sub branch of the Torpedo Branch. Telegraphist ratings were to be under the supervision of the Torpedo Officer and, when not working on WT duties, they were to work with torpedomen on other electrical duties.

At the outbreak of war in 1914 it became apparent that the limited instructional capability in radio frequency technology and WT equipment at *Vernon* was insufficient to properly support the Fleet. This led to a recruitment drive to bring suitably qualified scientists and engineers into the dockyards and ships to help either as civilian engineers or as RNVR electrical officers.[74] The RNVR officers were often appointed to ships to assist the Torpedo Officer who was, by now, expected to conduct his underwater warfare responsibilities along with his responsibilities for an ever expanding portfolio of electrical technology. The Torpedo Officer's problem was further eased in 1917 when research and development of WT equipment was transferred from *Vernon* to the RN Signal School and the Wireless Telegraphy Branch was merged with the Signals Branch. With this change, the Signals Officer took over the responsibility for shipboard WT operation, but not maintenance, from the Torpedo Officer.

Because of his professional standing before entering the service, the RNVR electrical officer was accorded wardroom status. This set a precedent for a commissioned electrical presence on board ship for the first time, as there is no evidence of any Commissioned Electricians then being in existence. Unfortunately, because warrant officers did not have wardroom status, the precedent accorded to the RNVR officers added to the dissatisfaction felt by many with regard to the standing of the warrant officer rank in that it was seen as isolated somewhere between the lower deck and the wardroom.

Support for Asdic

In 1917 the Anti-submarine Department of the Admiralty had set up an anti-submarine warfare training school based at HMS *Sarepta*, the naval air station at Portland. In anticipation of the imminent arrival of asdic, a training course for asdic officers was subsequently established at Dartmouth in August 1918. As part of the course, these officers were also sent to Parkeston Quay in order to receive instruction in the operation and the maintenance of the new equipment. However, the ship fit programme stalled at the time of the Armistice and the seagoing requirement for the asdic officer was delayed until around 1922.

Sarepta was closed down as an air station after the First World War, but the Portland-based AS instruction facility was kept open with sea training being carried out in the 1st Anti-submarine Flotilla, which comprised a number of P and PC Class patrol boats and some supporting drifters which had been retained as anti-submarine support vessels.

In 1920, the Field/Waistell Committee – which had been investigating ways of relieving the Torpedo Officer of his technical workload – also recommended that all aspects of anti-submarine warfare should come under the control of one branch. These findings may have contributed to the Portland AS school being reopened in 1920 and an Order in Council, CW5187/1920 dated 11 March 1920, was issued to establish the Military Branch Anti-submarine (AS) specialist officer's allowance of 2s 6d per day. This was followed by Order N12856.1921 dated 10 August 1921 introducing the non-substantive rate of Submarine Detector (SD) Instructors to provide the operator and maintenance rating support for asdic. By the end of 1921, the embryo of an anti-submarine branch was in being with seven specialist AS officers, three instructors and 37 SD first and second class ratings forming the core. In 1922, the Admiralty Board formally accepted the recommendations for an Anti-submarine Warfare (AS) Branch and announced its creation in AFO 1815/22. In 1924, the AS school became an independent command and was commissioned as HMS *Osprey*, where the Anti-submarine Experimental Establishment was eventually collocated in 1927.

Post-First World War Restructuring

After the First World War, there was much concern that the Torpedo Officer was increasingly being diverted from his primary warfare role and having to devote too much of his time to the maintenance of complex electrical installations afloat. Apart from the burgeoning number of general electrical systems, this included the maintenance of the recently invented asdic. During the 1920s, various committees were formed to study how branch structures could be reorganised to provide satisfactory electrical support for the Fleet and alleviate the Torpedo Officer of some, if not all, of his technical duties.

In 1920, the Field/Waistell Committee concluded that a separate specialist Electrical Branch should be formed to assume responsibility for all electrical systems. In order to form such a branch, the committee recommended the introduction of a specialist electrical engineering officer who would take over departmental responsibility for the existing EAs and a newly instituted 'specialist electrical technical rate' which would replace the LTO rate. The main difference in duties for the new rate would be that the billet should not have any involvement with the executive or operational responsibilities of the seaman branches. It was felt that these recommendations, along with the formation of an AS Branch, would relieve the Torpedo Branch of its anti-submarine warfare and electrical duties and allow it to concentrate on its underwater weaponry role.

On 14 March 1921, another committee, under the presidency of Admiral Sir Frederick Tudor KCB KCMG and supported by Engineer Vice Admiral Sir George Goodwin and Engineer Rear Admiral Toop, was appointed by the Admiralty to look at the Engineer Branch taking on the responsibility for electrical equipment. Part of the committee's remit was to work out a scheme of training that would allow electrical specialist officers to take over the electrical maintenance duties of the Torpedo Officer in ships and to become eligible for selection to the higher electrical positions in the Admiralty and the dockyards. However, the Tudor Committee's terms of reference did not specifically include the investigation of a new branch structure.

In establishing the context of the study, the committee took into consideration the small number of volunteers for the Engineer Branch[75] coming through the Selborne-Fisher officer training scheme. It noted that it was

> … almost entirely due to the prevailing views as to the difference in the conditions of service in the Executive Branch and the Engineer Branch. The younger officers of the service evidently think that the outlook of the Executive Officer is more pleasant while his prospects, at least in the higher ranks are probably considered to be more alluring than those for an officer in the Engineer Branch. The officer who does not volunteer for (E) retains military command and has before him the possibility of attaining the highest rank in the service, which the officer who volunteers for (E) does not have.

The Tudor Committee recommendations fully endorsed the concept of specialist marine and electrical engineer officers and made detailed proposals for the modification of facilities at Royal Naval Engineering College, Keyham to cater for an increase in electrical training. Although reference was made to separate marine and electrical engineering branches, it was considered that on board ship the two branches should be under the control of a senior engineer officer. The report went as far as recommending 74 posts which should be filled by electrical officers, including four appointments in the Admiralty.

While the matter of the shortfalls in engineer officer career expectations were excluded by the terms of reference, the committee did make some observations which included:

> We consider that Marine Engineer and Electrical Engineer Officers should not only be considered equally with Executive Officers for appointments as Admirals Superintendent, but also that, if possible, the command of all engineering training establishments should be given to them in lieu of Executive Officers as appointments fall vacant. The fact that high appointments in the Naval Service could be held by either branch should tend, in our opinion, to increase the esprit de corps of the service as a whole.

In 1922, the Admiralty Board, having accepted the recommendations for an AS Branch, turned down the recommendations for the Electrical Branch on the grounds of 'possible difficulties with regard to ship complementing'.

The recommendations for a separate Electrical Branch were reiterated by yet another sub-committee of the Admiralty Board in 1924. In order to overcome the problem of ship complementing identified in the Field/Waistell proposals, this committee proposed that the seaman torpedoman should have a common reporting chain into both the Torpedo Officer and a newly established Electrical Officer for the appropriate aspects of his duties. This committee also proposed that naval personnel should replace the civilians manning the dockyard electrical departments with the view of increasing the pool of uniformed electrical expertise. Although the Board is understood to have approved these recommendations, the changes to the Royal Dockyards and the full concept of an Electrical Branch were never implemented. However, in apparent recognition of the views of all earlier reports, the 1925 Order in Council, which introduced a completely new branch structure, did include an Electrical Branch but it was to comprise the existing EAs and Warrant Officer Electricians. It did not address the need for a specialist commissioned Electrical Officer structure.

In 1926, the AS Branch took over responsibility for the operation and maintenance of all asdic anti-submarine equipment and the workload of the Torpedo Officer was reduced accordingly. In 1927, the Anti-submarine Experimental Establishment was formed at *Osprey*, Portland from staff drawn from the asdic division of the Signal School at Portsmouth, but the anomaly of the Signal Branch being responsible for the development of asdic submarine detection equipment remained until just after the start of the Second World War.

In 1928, it was proposed that the ship's Engineer Department should, once again, take over all high-power electrical maintenance from the Torpedo Department. This proposal was modified in 1929 and the Engineer Department assumed responsibility only for those high power systems of which they were the sole users. Thus in 1930, the department held responsibility for the maintenance and operation of all high power installations within the department's own machinery spaces, with the exception of the electrical end of the turbo-generators which stayed with the Torpedo Officer and the EA. The result of this demarcation was that by the beginning of the 1930s the Engineer Department had assumed responsibility for only a small portion of high power systems. This was owing to the fact that the Signals and AS departments had retained responsibility for the maintenance of their departmental equipments from the high power supply fuses. By this time, nearly all ship departments were becoming users of general electrical equipment and habitability improvements were causing more complex compartment ventilation and lighting systems to be installed. This meant that as fast as the Torpedo Officer was relieved of some maintenance burden, new general ship services equipment was being fitted to take its place.

The failure of the Admiralty to implement the early recommendations for a separate Officers' Electrical Branch and the continuing growth of non weapons-related electrical systems, also meant that the electrical maintenance workload of the Torpedo Department became heavily biased towards general electrical equipment rather than torpedoes, mines and other underwater ordnance. In response to the evolution of what were effectively two diversifying fields of technology, in 1935 the Torpedo Branch introduced the Whitehead and Electrical specialist qualifications (SQ) at the level of Torpedo Gunner's Mate. The Whitehead SQ retained responsibility for the torpedoes and underwater weapons and the Electrical SQ took on the maintenance of low power switchboard and distribution systems including internal communications. By 1938, the LTO became even more specialised by being trained in either fire control systems associated with torpedo and gunnery systems or low power electrical systems, with the latter specialist being designated as the 'low power LTO' (LTO(LP)). Throughout this period the EA continued to work for the Torpedo Officer as a skilled electrical technician on gyro, high-power electrical machinery or torpedo maintenance.

CHAPTER 4

WARTIME DEVELOPMENTS 1939–1945

Communications

The naval communications system had been rapidly expanding during the build up to war and it was eventually operating on a worldwide basis. Fixed service communications had been established between shore-based communications centres using HF sky wave transmissions and directional aerial arrays. This gave global communication coverage but it was subject to fading and propagation loss due to the changing conditions in the ionosphere. Space diversity[76] and frequency diversity were used to maximise coverage under all atmospheric conditions. These techniques were supported by a VLF transmission capability which enabled a ground wave propagation path around the world that was much less dependent on the condition of the ionosphere.

A number of WT support ships were also brought into service and these ships had a fixed service capability, similar to the shore stations. They were used as deployable units and able to cover any communications 'black holes' encountered in operational theatres. The main drawback with these ships was that the physical size of the optimum HF aerial requirements precluded use of fully matched aerial arrays and the use of space diversity techniques on board. While a degree of shortfall in receiver performance could be mitigated by moving the ship within theatre, some WT ships were fitted with commercial equipment capable of receiving the same transmitted information at three different frequencies and combining the detected output signals for conversion to tone signals and, hence, Morse code.

Shore-to-ship communication employed a broadcast principle with the shore station using omni-directional HF transmissions, unrestricted by power supply and space requirements, to reach the Fleet around the world. Transmissions would take place on a number of frequencies with ships listening out on the optimum frequency for their area. The shore stations also had the VLF capability for use at certain times of day when the HF sky wave path was uncertain.

Ship-to- shore and ship-to-ship communication, unless both parties had line of sight, was also conducted on either MF or HF frequencies. A major wartime concern was inadvertently giving away a ship's position to the enemy, and any form of radio transmission was potentially going to do exactly that, with the ether being actively monitored by both sides for any communications traffic. Using MF frequencies at low powers was one of the ship to ship preferred options as there was a smaller risk of any associated sky wave propagation and the ground wave could be limited, not just by the low transmitter power but also a degree of atmospheric absorption at the higher MF frequencies. Alternatively, the use of HF for ship-to-ship communications was less predictable. While the ground wave was reliable enough for just over the horizon coverage, even at low transmitter powers there was always uncertainty as to how far the sky wave at any particular frequency would travel.

Aerials were a particular problem at MF and HF frequencies for the warships. In the early days of the war, ships used long wire aerials stretching between masts for MF or shorter, vertical wires run down from mastheads for HF. Wire aerials were inefficient and needed a lot of power with up to 90 per cent transmission loss in the feeder before 10 per cent of the power was successfully radiated from the exposed wire aerial. From 1941, the capability of using MF and HF circuits for voice, rather than Morse key communications started to be introduced into ship-to-ship communications, but this was always accompanied by the risk of propagation beyond the ship Task Group under certain conditions.

One significant technical advance made during the war that made HF voice communications more reliable was the introduction of crystal control of transmitter frequencies. This technology provided more stable transmission frequencies which could be preset at the receivers and easily changed to meet the operational communications plan. One disadvantage was that an individual crystal was required for each frequency. This inflexibility was later overcome using a Variable Frequency Oscillator (VFO) output mixed with a crystal output to generate a full range of frequencies based on the fundamental frequency, or a harmonic, of the crystal. This provided the inherent frequency stability of the crystal without the need for specialist environmentally controlled enclosures.

Very soon into hostilities, RAF Fighter Command found that they had a requirement for a much higher number of frequency-separated voice channels for the control of aircraft defending the homeland during the Battle of Britain. A VHF communications system covering the 100–124Mhz band was quickly developed and provided a solution which not only gave the required number of channels but also reduced the interference problems which were becoming prevalent due to the wider deployment of radars operating in the HF band.

The Royal Navy tended to require fewer channels of communication for Force operations and seemed content with using wireless telegraphy (Morse), rather than radio telegraphy (voice), as it was less prone to distortion in the presence of HF radar transmissions. One reason for this lack of distortion was that the development of radio interference suppression (RIS) equipment which used the transmission trigger pulse in a radar system to generate a suppressing voltage in the amplification chain of the WT receivers and shut down communications reception during the instant of radar transmission. This technique worked reasonably well for wireless telegraphy but was less effective for voice owing to the wider bandwidth needed to carry a full audio signal. Various techniques evolved for interference suppression throughout the war as, particularly, radar frequencies started to move farther into, and higher than, the VHF band.

In addition to the use of RIS equipment, the siting of respective aerials was considered very carefully in order to minimise the potential interference effect. Unfortunately, during the early years of radar, interference was not confined to the area adjacent to the aerial. The feeder cables used to supply the aerials were also prone to signal leakage, made worse as transmission powers increased, and this was not properly resolved until the development of metal waveguide systems which retained the radar energy within the aerial run and prevented interference being picked up between decks.

Under these technical and operational circumstances, the Royal Navy did not give any priority to the use of voice communications until, as the war progressed, more combined operations with the RAF began to take place. This type of requirement became evident during the Dunkirk evacuation and its importance was further emphasised by the need for convoy protection using air cover on the approaches to mainland Europe. RAF VHF equipment started to be fitted in small ship escorts in order to communicate with joining aircraft and coordinate the anti-submarine operations which did so much to reverse the British fortunes during the Battle of the Atlantic. As more aircraft were deployed overseas, so this type of tactical communications capability became even more important.

In hindsight, the Royal Navy's late adoption of VHF was difficult to understand as there were significant advantages associated with using VHF frequencies. Firstly, the propagation characteristics of VHF meant that transmission ranges were, theoretically, limited to the horizon due to the transmission being quickly attenuated by the atmosphere. This offered a degree of communications security not reliably available using low power HF transmissions. Secondly, and perhaps more importantly, the size of the dipole aerial required was 100 times smaller than that for an MF aerial and 10 times smaller than for an HF aerial. Thus, introducing a VHF capability would, at a stroke, have reduced the need for HF and MF aerials, except those required to support ship-to-shore or long range ship-to-ship communications, freeing up valuable masthead and deck space for other purposes.

In 1942, it was decided to fit all major Fleet units – including destroyers and 50 per cent of corvettes – with high and low power VHF transmitter receivers in preparation for the campaign in North Africa. At around this time, the Fleet Air Arm also adopted the use of VHF communications for fighter control. This change greatly extended the requirement to fit the British manufactured equipment, already in use by the RAF, into the Fleet, and the extra demand caused shortages in the supply chain. The shortfall was initially met by procuring US equipment, but this solution brought its own problems. One of these was the fact that the US equipment used frequency modulation (FM) techniques[77] in order to voice modulate the frequency of the carrier transmission wave, rather than its amplitude as was the case in British equipment. The use of the US equipment introduced some operating compatibility problems and affected the provision of support at sea in terms of spares and the training of personnel.

As a more permanent solution, it was eventually decided to take up dormant RAF VHF development work and use it to produce a ship-optimised equipment, Type 87M, which met the combined needs of the RAF and Fleet Air Arm. This equipment still utilised amplitude modulation (AM) technology and was designed to support five or more simultaneous channels of operation. Although the design aims were met, more cases of cross modulation between adjacent channels were experienced due to aerial constraints, particularly in aircraft carriers where aerial sites were confined to the island area.

In 1939, the method of communicating information over an HF link was still confined to the use of Morse keying at hand speeds and, later in the war, higher machine generated speeds. However, Morse was fundamentally unsuitable for use with the automatic telegraphy (AT) technology that was becoming the standard means of communication over commercial land lines. AT involved teleprinter to teleprinter communication using a continuously transmitted binary signal made up from five bit coded elements of information known as the Murray Code, instead of the interrupted transmission of the two elements, dot or dash, used in Morse Code. As the war progressed so this alternative to Morse was adopted for UK military service and AT communication became the standard for a growing proportion of the Fixed Service signal traffic. The use of AT for ships was not brought into use for warship communications until 1946, when a system was fitted to HMS *Vanguard* for the ship's royal tour to South Africa. The AT capability was not to be fitted Fleet-wide until the mid 1950s.

Direction Finding

During the Battle of the Atlantic, German U-boat transmissions were being intercepted by HFDF stations ashore but, while useful for early warning of submarine wolf packs gathering to meet Allied convoys, the information was not good enough to fix the submarine position for tactical use at sea. Unfortunately, the poor performance of the existing ship HFDF systems during the early stages of the war, even though fitted in prime sites at the masthead, had done little to overcome this tactical weakness. Consequently, convoy escorts only had asdic available for submarine location at a range where it was, arguably, too late to avoid convoy losses. Apart from the range limitation, asdic coverage was also limited in its search capability owing to a narrow acoustic beamwidth; and it was only really effective against submerged submarines.

In mid 1941, the Fleet's surfaced submarine detection capability improved with the introduction of the Type 286 (200MHz) and the Type 271 (3000MHz) radars. These radars were physically more suitable for small ship fitting and, in particular, the Type 271 provided a surface detection and location capability for 360° about the ship and out to a range of 5000 yards for a submarine target. The consequence of this was that the sensor competition for prime sites was suddenly being lost by HFDF and aerials were being moved or removed to allow the fitting of radar aerials.

Notwithstanding the advance in radar, since early 1941 it had been realised that U-boat operating tactics depended on being able to communicate over HF. Consequently, HFDF was considered to be a more useful resource for gaining tactical advantage because of the potentially greater range of detection over a ship-borne radar. Accordingly, the Naval Staff raised the priority for an improved

FH4 aerial testing at HMS *Mercury*. (Collingwood Museum)

ship-borne HFDF equipment and, in 1941, this led to W. Struszynski, a Polish engineer based at the Admiralty Signal Establishment (ASE),[78] devising a 'Sense' aerial which, combined with the existing HF detection aerials, resolved the bearing ambiguity problem that had been the bane of pre-war HFDF work. Armed with accurate bearing information, ship escorts were now able to obtain Force-based submarine intercepts using triangulation. This gave increased tactical range advantage over both the Type 271 and the Type 286 radars. Although the convoy escorts were reluctant to downgrade their radar capability by moving the aerials, the priority of HFDF fitting was restored by the Naval Staff in late 1941 and the FH3 HFDF equipment started to be installed in the Fleet.

FH3 comprised a masthead aerial, often known as the 'Birdcage', which incorporated a sense element and a structural design which enabled components of received sky wave transmissions to be nullified, thus reducing the errors caused by the varying angle of incidence of the sky wave at the aerial. This system gave a positive identification of the true target bearing and started to regain the trust of the Fleet in HFDF. FH3 was followed in 1943 by the introduction of FH4, still with the ubiquitous Birdcage aerial but now equipped with a much more effective operator display interface.

FH4 offered significant improvements in the speed of gaining intercept information, given that U-boat transmissions were of a very short duration. Not only was the bearing information more accurate and timely, the equipment allowed a skilled operator to tell whether the intercept was from the sky wave or the ground wave element of the transmission. This was tactically important as the ground wave transmission was more reliable and would have been from a much closer submarine and, therefore, more of a threat. In addition to direction finding, FH4 was also capable of measuring the received radio signal field strength at the ship and, based on U-boat known transmitter power, calibrated curves could be used to estimate the range of a source transmitter.

It was from this time that HFDF started to make its major contribution to the Battle of the Atlantic. The 'Birdcage' aerial continued to remain a common sight in the Fleet until the 1980s.

Radar

In 1939, only two operational ships could boast what the Royal Navy was calling RDF but the RAF was starting to know as radar. Those ships, *Rodney* and *Sheffield,* were fitted with the Type 79Y radar, a system which operated at around 40MHz with a wavelength of 7.5m. Installation of 40 production Type 79Ys had started in January 1940. An improved Type 79Z became available in May 1940 but, like its predecessor, it was only suitable for fitting in cruisers and above. Both variants were soon found to meet the requirement for general air surveillance but poor range and bearing accuracy did not make them suitable for fire control purposes.

Anxious to harness the potential of radar for range finding and replace the existing optical rangefinders being used in both air and surface warfare roles, the Naval Staff pressed for a capability to be added to the Type 79Z. This was despite the fact that, in February 1938, it had

HMS *Warspite* with RDF Type 79Y fitted. (Collingwood Museum)

HMS *Sheffield* with Type 79Y RDF aerials, 1942. (Collingwood Museum)

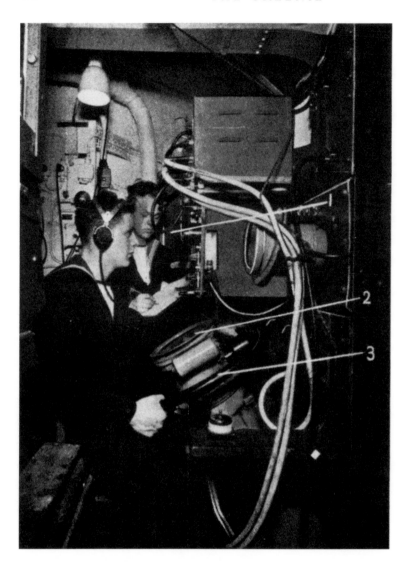

RDF Type 279 radar office. (Collingwood Museum)

been accepted that an accurate fire control capability would only be achieved by the use of much higher operating frequencies. In the absence of a higher frequency option, the solution offered by the Signal School was to integrate the Type 79 radar with an Army-designed ranging unit that could display precision range and bearing using two A scan cathode ray tubes. The combined equipment was designated the Type 279.

The Type 279 controls incorporated a short radar pulse facility and, when the target return signal was displayed separately on an expanded range scale, this allowed more accurate range interpolation of the received radar data. Unfortunately, the radar still only had the inherent bearing accuracy of the original Type 79Z and so the existing optical systems were still required in order to obtain best bearing information. The other weakness of the Type 279 design was that when switched to short pulse the radar could not be used for air surveillance and this was often operationally unacceptable. Another shortcoming, common to both the Type 79 and Type 279, was the need to manually rotate the separate transmission and reception aerials. This coordination requirement was itself difficult in anything other than calm weather and wasteful of prime masthead positions at a time when the DF equipment coming into service also needed such a position for optimum performance.[79] However, probably the most

Chain Home radar control room *c.*1940. (BAE Systems Archives)

serious limitation was that none of the radar sets in service could yet be fitted in ships smaller than a cruiser.

Even by the end of 1939, it had been accepted that the required surface detection and fire control capability would not be met by the proposed Type 279. Therefore, in the absence of significant progress on a 600MHz fire control radar and with hostilities escalating, it was decided to develop a radar using newly developed silica valve technology. This technology was found capable of increasing the transmitted power available to almost 1000kW but only at a marginally higher frequency of 85MHz (3.5m wavelength) than the 40MHz (7.5m wavelength) used in the Type 279. In theory, this frequency offered only limited benefits of reduced aerial size and improved surface detection and ranging performance but there were no other immediately viable options. Accordingly, it was decided to develop a multi-purpose radar which could be used to fulfil all roles to a limited extent and would be more suitable for fitting in smaller ships.

The 85MHz equipment was given the designation Type 281. It used the same twin aerial system as the Type 279 but rotation was achieved by an electric motor that automatically reversed the direction of aerial rotation before it traversed through the ship's stern arc.[80] Shore development trials started in May 1940 and the prototype Type 281 performed well. Fortuitously, the radar was

WAAF Chain Home radar
operator. (BAE Systems Archives)

in position and able to detect and track the first German air raid over Portsmouth on 11 July 1940. This in itself was a significant event as the RAF Chain Home system of warning radars in south-east England did not at the time extend as far west as Portsmouth. As a temporary measure, the Type 281 was retained in operation at Eastney until the RAF radar coverage had been extended farther along the south coast.[81]

In the light of the early shore-based success, the Admiralty ordered 36 Type 281 production sets in June 1940 and released the prototype for sea trials in HMS *Dido* in August 1940. During the trials, aircraft at 5000 feet were detected at 60 miles rising to 110 miles at 20,000 feet. Surface detections were achieved at 11 miles on large ships and this was regarded as a success. Aircraft trials did highlight that the higher transmission frequency caused the transmission lobes to be lower in height than for the Type 79 and this reduced the maximum height of aircraft detection from 40,000 feet down to 28,000 feet. Nevertheless, the first production Type 281 radars started to reach the Fleet towards the end of 1940, and by late 1941 a single aerial working variant was also introduced.

While Type 281 went some way towards meeting the surface ranging requirement, it did not help with anti-aircraft fire control, which required continuous range to be fed to the HACS system for the prediction of target future position allowing for the shell's time of flight. The urgency of this requirement was brought home during the Norwegian campaign in May 1940, when the Royal Navy suffered considerable losses because of the poor performance of the HACS optical coincidence rangefinder against enemy aircraft which were being tracked in by Type 279 from ranges out to 40 miles. The Naval Staff's preferred fire control radar system was one that could be co-located with each gun battery, thus avoiding the bearing parallax problem, something the Germans had implemented on their main anti-surface armament as early as 1938. By providing continuous range information, ammunition fuzes could also be set for a range at which to establish a barrage, a tactic thought particularly useful for both close-range weapons, such as pom-pom batteries, and main armament batteries in deterring dive bomber attacks.

Trials using a concept 600MHz radar, with a transmitter power of only 15kW, had been carried out in February 1940 and shown that low flying aircraft could be detected and tracked in from 5000 yards down to a few hundred yards. This demonstration convinced the Admiralty to place a verbal order for 200 sets, designated Type 282, some five months before system design would be completed. Sea trials were carried out in HMS *Nelson* using a prototype mounted on a pom-pom director in June 1940 with convoys being detected at 36,000 yards and a destroyer at 23,000 yards.[82] The Admiralty's reaction was to order a further 700 sets and authorise fitting in destroyers and above. The Type 282 design quickly evolved into three further variants, Types 283/284/285, for use in the control of close-range batteries and main armament in both anti-aircraft and anti-surface modes. While the speed of the 600MHz development programme and delivery to ships was to be admired, there were some distinct limitations to the radar design. These included low transmission power with a resultant limited range, bearing accuracy not being good enough for blind fire, mutual interference in ships with multiple installations and component reliability. It was decided that a programme of improvements should be established to alleviate some of the limitations and that any modifications to the existing equipments should be carried out at the earliest opportunity.

This policy reflected the general approach to radar system improvements simply because the time was not available to properly engineer a seagoing system from a prototype to a robust and maintainable design, as the need for some capability in the Fleet was too great.

Following the Dunkirk evacuation in June 1940, the requirement for a surface detection radar capable of installation in small ships became paramount under the threat of invasion. The Signal School identified a possible solution using an RAF air-to-surface surveillance radar, ASV Mk1, which had acceptable characteristics in that it operated at 214MHz giving a wavelength of 1.5m. The concept was tested out in 1940 using a Walrus seaplane placed on the slipway at Lee-on-the-Solent, Hampshire, and surface targets were detected in the Solent at around 10,000 yards range. Full adaptation was started and a naval version, designated Type 286, was fitted for trials in HMS *Ambuscade*, a destroyer of some 1200 tons. The radar was deemed sufficiently successful to start production fits into other small ships of the Fleet starting as early as June 1940. The main drawback with the Type 286 was that the aerial was fixed and the ship had to be pointed at the target to obtain accurate bearings. This problem was not alleviated until a rotating aerial was developed and replaced the fixed aerial installations in 1941.

As the threat of invasion receded, so the Battle of the Atlantic started with a vengeance and the need for a high definition surveillance radar that could detect a submarine on the surface became of the utmost priority. A radar operating at 3,000MHz (10cm wavelength) had been predicted as offering this type of performance and, additionally, it would meet the small ship fitting requirements as well as minimise the chances of being detected by enemy DF equipment outside horizon range. It was not until early in 1940 that technological advances were made which finally brought the centimetric radar into being. The first breakthrough was the invention of the multi-cavity magnetron valve, which could produce sufficient transmitted power at the required higher frequency. This valve, coupled with the newly developed reflex klystron valve capable of producing similar frequencies for the radar receiver function, formed the basis of a shore based 3000MHz prototype radar, able to detect HMS/M *Usk* on the surface at 7.5 miles in November 1940.

This first breakthrough was soon followed by the development of a duplex switching system that allowed radar pulse transmission and reception through the same antenna. This technique not only resolved more space and weight concerns but it also solved the problem of synchronising the transmitter and receiver bearing lines. With this critical technology now made available, the physical size reductions achievable at the new centimetric frequencies opened up boundless possibilities for fitting minor war vessels and submarines with surveillance radar.

In March 1941, the sea trials of the Royal Navy's first centimetric radar, Type 271 (3000MHz and 10cm wavelength), took place in HMS *Orchis*, a Flower Class corvette. The height of the aerial was 36 feet and in flat, calm conditions a fully surfaced submarine was detected at 5000 yards. These results were well in excess of the visual sighting range of a submarine expected from a ship at night but considered technically disappointing when compared with shore-based trials using *Usk*, which was actually a smaller submarine than the trials submarine used by the *Orchis*. Nevertheless, by the end of the year some 1350 sets had been produced or put on order. The basic design of the Type 271 evolved into three variants, Types 271/272/273, for installation in various types of ship, and modified systems were produced for use ashore as part of the coastal defence chain. It was only after the introduction of the small ship radar that the Battle of the Atlantic started to swing in the Allies' favour. Even more fitting possibilities looked achievable in 1943 following the development of a 9000MHz, 3cm radar, which finally brought reductions in top weight and electrical power requirements to the point where high definition radar in the smallest ships and in aircraft was to become a reality.

An early spin off from the wartime advances in radar technology was the concept of Identification Friend or Foe (IFF) systems. This type of system involved friendly aircraft being fitted with a transponder, a small radar transmitter activated when the aircraft was in the swept arc of a friendly ship or shore-based radar beam. The output from the transponder was received as an echo by the triggering radar for coincidental display with the raw radar return from the aircraft. Enemy aircraft not so fitted were therefore identifiable as being hostile through the lack of a transponding echo being shown on the radar display.

As the number of radar frequencies increased, it became impossible to fit transponders to cover all possible radars and the 'interrogator' was introduced. The interrogator was a separate radar transmitter, synchronised to a primary surveillance radar transmission but on a dedicated frequency, and this transmission would generate the response from the transponder as an independent radar return. IFF systems continued to evolve and, in due course, the interrogation pulse and responding echoes could be coded to make the identification process even more selective. Apart from improving the tactical management of the air battle, IFF played a vital role in reducing the number of 'friendly fire' incidents.

Such was the high rate of progress in the radar field that the concept of a standard production set was almost unknown. As the research scientists came up with new ideas, so teams of radar engineers were continually travelling within the Fleet installing new equipment and modifying equipment already in the field. The inevitable problem was reliability, with most equipment and

(Collingwood Magazine Archives)

(Collingwood Magazine Archives)

modifications having been conceived ashore and not being engineered to suit the more rugged environment at sea. Apart from this design weakness, the situation was aggravated by the fact that when war broke out most ships had limited dedicated electrical expertise on board and there was no radar expertise. During the earlier war years this, and the equipment's Most Secret security classification, would have hampered attempts to get informed feedback from the sea, even if the host ships were able to remain in communication. One aspect of feedback which would have been prevalent was the new problem of mutual interference between radar sets when transmitting simultaneously from within the relatively small area of a ship. The installation and maintenance standards used would have been a big enough problem on their own; but developing and implementing the concepts of pulse synchronisation and receiver blanking to reduce mutual interference was something which needed a total systems engineering approach that was almost impossible under wartime conditions.

Given these constraints, it seems astonishing that, in the face of what was probably the most rapid advancement of a new and complex technology ever, by the end of the war most ships were deploying some form of radar capability. In fact, by 1945, almost 100 radar types of various mark and modification numbers had been produced for the Fleet to meet some naval warfare requirement or other.[83] A yardstick against which to measure the magnitude of this achievement can be gained by comparing the single metric radar fitted in two capital ships at the beginning of the war, to the typical battleship or cruiser radar outfit as fitted around 1944.

Typical Battleship or Cruiser Radar Outfit c. 1944

Radar Type	No. of Sets	Function	Frequency
Type 281B	1	Long range air warning	86–94Mhz~
Type 293	1	Weapon control and target indication	3000Mhz~
Type 277	1	Surface warning and height finding	3000Mhz~
Type 275	2	Anti-air armament control	3530Mhz~
Type 274	1	Surface main armament fire control	3300Mhz~
Type 253P	1	Interrogator	157–187Mhz swept frequency
Type 251M	1	Transponder for use with RAF	176Mhz~
Type 941	1	Air contact interrogator	209Mhz~
Type 242WS/ WC/GC	3	Surface, air and fire control contact interrogator	179/182Mhz~
Type 243	1	Air contact interrogator	171/179Mhz~

Typical radar outfit fitted in battleships and cruisers *c.*1944. (Collingwood Museum)

The rate of development is also demonstrated by the changes in equipment when comparing the eighteen radar sets fitted in HMS *Vanguard* when she completed build in April 1946, to the radar types of typical battleships less than two years previously. A noteworthy point is the integration of the interrogation function with a parent radar.

HMS Vanguard Radar Outfit April 1946

Radar Type	No. of Sets	Function	MHz
Type 960	1	Long range aircraft warning with IFF interrogation	88
Type 277	1	Surface warning and height finding	3000
Type 268	1	Surface warning	9400
Type 293P	1	Target indication with IFF interrogation	3000
Type 275	4	Anti-air armament control (forward and aft batteries)	3530
Type 274	2	Main armament fire control (forward and aft batteries)	3300
Type 930	1	Main armament fall of shot detection (Army variant)	3000
Type 262	6	Close range gunnery fire control	9650
Type 953	1	Transponder beacon	182(Tx) 9400(Rx)

Radar Counter Measures

One might conclude that the secrecy of the radar programme meant that the Allies were the only ones with radar, but the Germans also had the technology. Following the Battle of the River Plate in 1939, the German pocket battleship *Graf Spee* was found to have radars operating at 500MHz for use in gunnery range finding. The radar was co-located with the gun but the range was limited and more importantly, the bearing accuracy was poor and had to be established optically .

German capability was most obvious when used for shore battery firings on British shipping traversing the straits of Dover, and this became very evident in the summer of 1940. Transit shipping was suffering high losses at night, believed to be due to accurate radar range-finding facilities linked to the German shore batteries. A Signal School team was sent to investigate and they detected radar transmissions around 83MHz coming from the other side of the Channel. Two sources of radar signal were identified and immediately a Radar Counter Measures (RCM) Group was formed to address the problem. This led to the development of a tunable jammer that could deny the Germans any range information. The system was tried in earnest in early 1941 and successfully caused the shutdown of the German radars and the safe passage of a convoy in transit at the time. RCM had been born and, by June 1941, the Type 91 jammer went into production and was being fitted to all battleships and cruisers by 1942.

Although the Admiralty was concerned about revealing the true level of British radar capability by the use of jammers, the development of radar jamming continued as the Germans developed counter-counter measures, such as re-tunable radars. A more critical use of the technology came when the Germans successfully used the radio-controlled glider bomb, HS293, at the Battle of Salerno in 1943. Initially, the Type 91 could not tune to an individual HS293 control frequency quickly enough but, as an interim solution, ships were ordered to use personal electric razors to superimpose broadband noise on the jammer signal and this, apparently, disrupted the radio control signal. The subsequent jammer, Type 671, was designed with a broadband noise jamming capability and the threat of the HS293, and its armour piercing variant, the FX1400, receded.

The next phase of wartime RCM was to introduce the concept of 'spoofing'. This was developed to confuse the enemy by radar transmissions from a notional decoy force which was intended to appear as the D-Day invasion force but approaching an area well north of the Normandy beaches. Selected German shore radars were allowed to continue operations to permit detection not only of these decoy radar transmissions but also ship echoes, which had been enhanced by the addition of 'window', strips of foil matched to the German radar wavelengths and which

German Wurzburg gunnery range finding radar. (BAE Systems Archives)

Dornier 217 carrying HS293 radio-controlled bomb. (Collingwood Museum)

gave radar returns indicative of major naval units approaching the enemy shore. Such tactics and technology represented only the beginning of this aspect of naval warfare.

Action Information Organisation

The problem of anti-aircraft defence for ships had been raised in a report by the Imperial Defence Committee in 1926 and this had led to the development of HACS, an optical fire control system based on existing anti-surface ship fire control computers. The HACS production equipment saw service during the early part of the war but the optical rangefinder proved to be completely inadequate against more modern attacking aircraft. The arrival of radar had appeared to offer better alternatives to visual detection and range finding and the application of this technology to air defence had been studied since around 1935 without much success until after war was declared. While development work eventually led to operational aircraft detection radars by 1939 and fire control radars by 1941, there seems to have been little effort put into the coordination of these capabilities in the conduct of naval warfare.

This shortfall was first brought home to the Royal Navy during what was probably the first radar-supported skirmish with the Germans on 26 September 1939. The *Rodney*, fitted with radar Type 79Y, was with the Home Fleet off Scapa Flow when she detected enemy aircraft inbound on her radar and, while tracking the aircraft in, she also reported the presence of the German force to the Fleet by flag. The initial warning proved of little value because the rates of change in range and bearing of the aircraft were too great to be usefully passed by voice to own ship visual positions, let alone by flag to accompanying ships. This missing capability later became known as 'target indication' and it became a critical function for the transfer of target data to fire control radars. More importantly, defensive opportunity was also lost because there was no evidence of tactical thought being given to the use of ship-borne aircraft to take advantage of the initial detection of the bombers some 80 miles from the Home Fleet, which at the time was accompanied by the aircraft carrier *Ark Royal*. Arguably, this incident led to an increased awareness of the need to manage tactical information gained from radar and use it to support the concepts of layered defence, a fully integrated combat system and carrier-based aircraft direction

'Target Indication'
explained from
first principles!
c. 1944.
(Collingwood
Museum)

operations. In due course, the solution to this warfare management requirement was designated as the Action Information Organisation (AIO) function of naval warfare and it started a concept which was to evolve almost as relentlessly as the underlying technology during the subsequent years.

One major step forward in the wartime development of AIO within a single ship was the invention of the Plan Position Indicator (PPI) display. This display presented a 360° radar picture about the ship. The picture was centred on the ship but it was north stabilised such that the radar picture remained fixed to the eye of the observer. Any radar contacts appeared as bright spots on the screen along a line of true bearing measured at the periphery of the screen. Range was measured using illuminated concentric range rings calibrated to present the range from the ship. This method of representation was a significant breakthrough as it allowed the operational picture to be continuously displayed and quickly appreciated by personnel elsewhere in the ship and not just the dedicated radar or sonar equipment operator.[84] Thus, weapons groups could see and relate the buildup of enemy force movements to their own ship's course and firing arcs, without the need for voice updates, and the Command could quickly make decisions based on the relative threat of each attacking force. By mid 1943, the PPI was being widely fitted to the Fleet to support most types of warning radar.

While the concept of AIO was fully recognised, there were physical and technical problems with implementing a fully coherent system even in new ships. Attempts were made to co-locate radar operating functions ranging from area surveillance and height finding through to fire control and target identification, but this was only possible in a limited fashion in a few ships before the end of the war. It was to be the arrival of the digital computer that would eventually make AIO a fully achievable goal and underpin the rise of the fully integrated combat system.

Power Generation and Distribution

The U-boat threat during the Second World War quickly exposed the weaknesses in the design of ship's power distribution systems. *Ark Royal* was sunk in November 1941, hit by only one torpedo that resulted in the immediate flooding of a boiler room and the main switchboard control room. The electrical teams found that they could not remotely control the distribution system and hand operation was impossible because of the damaged state of the ship. It was evident that development of a new distribution and control system was of paramount importance and the outcome was a relay system in which all the control cables during normal operation were left without any power on them. The new system made it practicable to introduce a secondary control position in the ship which could be brought into operation if the main control position became unserviceable.

The loss of *Ark Royal* also emphasised the necessity for ships to have generators which could maintain essential electrical supplies after a loss of steam to the prime mover. Immediate steps were therefore put in hand to fit back-up diesel generators into all ships. At the same time, automatic changeover switches were introduced to enable alternative power supplies to be quickly provided to close range anti-aircraft gun mountings and other essential services. The Second World War experience also led to the development of improved damage control facilities including emergency communications systems and high integrity secondary lighting using battery-supplied and automatically switched emergency lights.

The change to AC had been investigated in 1883 and 1932, but the balance of opinion was then in favour of retaining DC. However, since the 1932 review there had a huge increase in power requirements; the electrical load for a typical cruiser had risen from 900kW to 4400kW. For the Royal Navy, the prospect of having to fit generators of at least 1000kW capacity – even in small ships – in the post-war era made a reappraisal of the position essential. It was clearly necessary to raise the supply voltage from 220 volts and to decide whether DC or AC should be used. Investigations showed that even for a small vessel with a load of 1000kW using a 440 volt, 3-phase system operating at 60 cycles per second could result in significant savings of weight and space, with those savings increasing with the size of the load. The maintenance load was also considered in the deliberations on this matter, as was arcing from DC machines which was recognised as a source of interference that could affect the performance of the new generation electronic

HMS *Ark Royal* would be sunk on 14 November 1941. (Naval Museum Portsmouth)

equipment. Finally, the experience of the US Navy since 1932 had shown that, if the electrical equipment was designed with safety in mind and the personnel were given suitable training, the use of 440 volts AC was not the significant hazard perceived by DEE in their 1932 review.[85]

The selection of the frequency for new AC systems was not easy, but it was evident that a motor speed of 1800 rpm met a large number of requirements and that the lower speeds, associated with a 50 cycles per second supply, would introduce an unacceptable weight penalty. Given the facts presented, and with the full support of industry, a frequency of 60 cycles per second was selected and this frequency is still used for the power generation systems currently in Royal Navy service.

The first ship with an AC system was HMS *Diamond*, a Daring Class destroyer, and this was followed by several classes of frigate. The Tiger Class cruisers were the first large ships to have an AC system but although these ships were laid down during the period 1939–45 they were not completed before the end of the war. The change to AC was timely as it was estimated that future electrical power requirements could only be met by replacing the original 500kW 220v DC generators with 1000kW 440v 3 phase AC generators.[86]

Although AC power generation systems were not in themselves new and equipment was available in the commercial market, the change raised many design problems. Auxiliary equipment, such as automatic voltage regulators, air-break switchgear, control gear, distribution equipment and emergency cables systems all needed to be made sturdier to meet naval environmental specifications, particularly in resistance to mechanical shock and water ingress. Trials were carried out in *Diamond* at the earliest opportunity in order to study system and switchgear behaviour under the often unique fault conditions which could be encountered on a warship. These trials helped to define the changes necessary to improve discrimination in fault protection systems and the survivability of the ships concerned.

Mine Countermeasures

In the Second World War the Royal Navy had to counter the German magnetic mine. This type of mine was able to detect a steel ship's magnetic signature and detonate when the ship was in close proximity. Such a mine had been available to the Royal Navy at the end of the First World War, but there appears to have been a marked reluctance to use it in the later conflict, possibly because there were stil no British countermeasures in place at the time. One of the first ships to suffer mining was HMS *Belfast* when she detonated a magnetic mine on 21 November 1939 in the Firth of Forth. The hull remained watertight but the shock damage to the machinery spaces and electrical equipment was so severe that the ship was immobilised and repairs were not completed until 1943. Although obviously devastating, this incident was forensically analysed with respect to the effects of the explosion throughout the ship. In addition to finding that the levels of shock were much higher than those previously being used to test ship's equipment, it was also established that cast iron machinery casings were liable to fracture as a result of underwater explosions. The information gained from the *Belfast* incident provided valuable data for subsequent ship design and escalated what had been a known threat up the priority chain.

Fortunately, in November 1939, a German air-dropped magnetic mine was found exposed at low tide on the beach at Shoeburyness. It was successfully recovered and its operating mechanisms investigated within the context of the considerable research knowledge already in place, and countermeasures were quickly brought forward. One of these was the development of the magnetic sweep to create a magnetic field that simulated the magnetic signature of a steel hull and caused the mine to detonate harmlessly under a towed, expendable body. The first sweeps were made using electrically energised coils wound onto wooden rafts to produce a simple simulation of a ship's magnetic field. The rafts were then towed at a safe distance behind a wooden towing vessel through the minefield thus sweeping it clear for any following steel ships. Another solution was the magnetic ship. This was a remarkable vessel that had a wire-wound magnet, about 100 feet long and weighing some 500 tons, fitted to the forward end. The magnet was energised from a DC generator, later converted to an AC supply of low frequency, to sweep mines of north and south polarity. Use of these ships was discontinued when magnetic mines were fitted with more sophisticated arming devices, which required an unknown number of actuations before exploding. In due course, this led to the development of the LL towed buoyant cable sweep which produced a large magnetic field in the sea by pulsing currents of 3000 amps through pairs of buoyant cables that were towed by minesweepers working in company.

It was not practical to have each steel ship accompanied by a minesweeper and this raised a requirement for a system to neutralise a ship's magnetic signature to effectively zero as a means of preventing detection by the mine sensors. The proposed solution was to fit the ship with degaussing coils – electrically energised coils that generated a magnetic field designed to nullify the ship's magnetic signature. Early degaussing installations consisted of coils sewn up in canvas and clipped outside the hulls. This system evolved into a single turn of rubber-insulated heavy copper strip fitted to the external plating. Following considerable research in 1940, it was found possible to obtain effective signature screening by fitting electrically energised coils inside the ship's plating and calibrating the magnetic field set up by the coils to offset the ship's induced magnetism, such that the resultant magnetic signature was reduced to undetectable levels.

Underwater Weapons

By 1939, the Royal Navy's torpedo capability in both ships and submarines had suffered because of a lack of stocks and a shortfall in production capability which had to be spread across a large number of types of torpedo rushed into service at the start of hostilities. Most of these torpedoes used a burner cycle combustion engine, a development of Whitehead's compressed air engine which used burning fuel to raise the air temperature before its introduction into the combustion chamber and fuel injection prior to ignition. Known as a 'thermal' engine, it could propel a 21-inch torpedo at around 50 knots. This type of engine was used in the Mk8, Mk9 and Mk12 torpedoes, fired from submarines, ships and aircraft respectively, and formed the bulk of Royal Navy expenditure during the war.

The pre-war advances in asdic technology had led the Royal Navy to investigate its application to the guidance of torpedoes. However, the thermal engine weapons in operational use were found to be too noisy above 20 knots for an acoustic homing system to be viable and the need to sacrifice speed for a homing capability was not seen as a priority. Although pieces of German electrically propelled torpedoes were captured in 1940, there was no urgency to copy the technology until the requirement for a trackless weapon for use in the Mediterranean was eventually raised. This led to pre-production, electrically driven weapons being available in 1943. However, by the time production models of the torpedo – designated the Mk11 – were made available to the Fleet, the naval campaign in the Mediterranean was over. Although attempts to get the weapons into the Japanese theatre were made, this did not happen until after the Japanese surrender and the British electric torpedo was never actually fired in anger. It is of interest that, although the Germans had deployed an operational homing torpedo by January 1943 – the Naval Acoustic Torpedo (GNAT), which accounted for the sinking of 25 naval vessels and 20 merchant ships during the war – a fully operational British version did not reach the Fleet until the arrival of the air launched Mk30 torpedo in 1954.

Despite the lessons learnt from the submarine threat in the First World War, little progress seems to have been made in the area of improving submarine kill capability between the wars. Depth charges and the associated ship launch systems did not evolve noticeably and, while asdic performance improved, the Royal Navy was still unable to determine submarine depth and therefore develop an accurate fire control solution. Perhaps more worryingly, there did not seem to be too much concern about this missing piece of target information, possibly owing to the diverse allocation of Headquarters' responsibility for the various components of an ASW system. This situation, plus peacetime financial constraints and not a little complacency about asdic operating limitations against submarines, may well have contributed to the low priority afforded to improving the capability to actually destroy a detected submarine.

Depth charge launcher fitted on Second World War Loch Class vessel. (Collingwood Museum)

In 1939, the continued use of amatol as the explosive fill for depth charges meant that the kill distance remained of the order of 20 feet. With such a limited kill range, the Royal Navy quickly found that the standard five depth charge pattern was inadequate and the number was boosted, first to ten and then to fourteen depth charges. This type of pattern was launched over the stern and either side of the ship using improved depth charge launchers. Early in the war, amatol was replaced by Minol, a more powerful explosive, and a heavier design of depth charge, the Mark VII, was introduced to reduce the sinking time. The Mk X depth charge, a much larger weapon containing one ton of explosive and the equivalent of ten standard depth charges, was quickly developed for use against the deep diving submarine but there is no evidence that this was operationally successful. The U-boat response was to move faster and dive deeper, and without any depth measuring capability the Royal Navy did not really know the extent of the problem on their hands. In fact it was not until June 1943 that confirmation was gained that U-boats could dive below 700 feet and this resulted in a staff requirement for a weapon capable of attacking a submarine at 1200 feet depth.[87]

The North Atlantic ASW environment exposed the limitations of the existing asdic equipment. Contact ranges were limited and usually predicated on intelligence reports based on Enigma-sourced intelligence and HFDF intercepts ashore being used to close the U-boat's position. When contact was eventually gained, the depth charge delivery systems were inadequate in that the submarine was lost to the asdic as the ship closed to pass over the target position and drop its depth charges astern. Even if depth information had been available, the lost contact time and the sinking time of the charge to depths down to 700 feet, when aggregated, gave the U-boat considerable time for evasive manoeuvres.

In order to overcome the tactical constraints of their asdic systems, Captain John Walker developed the 'creep attack' for use by his Second Escort Group. This involved two or more escort ships working together with one ship using its asdic to locate the target and the others running quietly with no asdic operating. When contact was achieved and held by the active ship, the 'quiet' ships were directed to drop depth charges on the known position and contact was never lost. The submarine, unaware of the 'quiet' ships, would have been waiting for the 'pinging'[88] ship to lose contact before starting its own evasive action. This tactic resulted in a much higher U-boat kill rate. Even so, the expenditure in depth charges was still significant and Walker's tactics were, on occasions, questioned by his operational commanders. However, it was the innovation of such tactics – along with the improvement in radar and HFDF capability – that started to turn the Battle of the Atlantic in the Allies' favour.

One solution to the tactical problem was a forward-throwing ASW weapon. This concept had been considered during the inter-war years but none of the research had come to fruition before the outbreak of hostilities. Apart from the number of Headquarters departments involved with trying to resolve the problem, progress was certainly hindered by indecision with regard to the nature of the weapon best suited to destroy a submarine. Some scientists preferred the use of larger charges because they required less accurate fire control data but such a system did impose considerable recoil forces on the ship's deck and relied on a pattern of explosions to crush the submarine. Others opted for lighter weight but more numerous charges spread over a wide pattern to compensate for poor target data and with less impact on ship's structure, which would allow greater ship fitting opportunities. There was also debate over the type of fuze to be used, either contact-initiated or with a preset depth; this was because the depth of the submarine was still undetectable at this stage of the war.

Prototype forward-throwing depth charge launcher known as the Five Wide Virgins and fitted in HMS *Whitehall* July 1941. (Imperial War Museum)

It was not until January 1940 that DASW authorised ASEE Fairlie to develop a triple-barrelled depth charge throwing system and, by this time, other Headquarters departments had started their own projects. These other systems included a rocket-fired projectile sponsored by the Propulsive Devices Establishment, a recoilless spigot mortar from the Miscellaneous Weapons Department and a stick mortar overseen by the Chief Superintendent of Dockyards. One of the first wartime systems to be actually trialled in a ship consisted of five mortars designed to fire existing types of depth charge over the bow in a fixed pattern. The system was produced by J.I. Thorneycroft, the Southampton-based ship builders, and became known as the Five Wide Virgins. A prototype was installed in HMS *Whitehall* and trials were held in the summer of 1941 but not considered a success. At the end of 1940, the Admiralty resolved the conflict amongst sponsor departments by nominating the Director of Naval Ordnance (DNO) as the responsible authority for producing an ASW system. A meeting was called in January 1941 to define the operational requirements which eventually led to the development of the Hedgehog anti-submarine mortar.

By the winter of 1942, an air launched version of the depth charge had been developed for the RAF but the opportunities for successfully deploying the weapon against a submarine were limited to visual sightings during daytime transit on the surface, which was invariably avoided by U-boat commanders. When aircraft radars were introduced which were capable of detecting U-boats in transit on the surface at night, unfortunately, the minimum range of the equipment was too great to allow the aircraft to retain contact until the depth charge drop position, in much the same way as ships would lose asdic contact. In mid 1942, the Leigh Light started to be fitted to ASW aircraft and this allowed radar contacts lost when closing to be visually re-established at night to complete the attack phase. This combination proved to be so effective that U-boats started to transit on the surface during the day when they had a better chance of sighting an approaching aircraft.

When the US entered the war in December 1941, the tactical situation started to change yet again with the introduction of airborne Magnetic Anomaly Detection (MAD) systems into the ASW campaign. MAD systems were used to detect the presence of a submarine underwater by sensing the localised distortion effect of the U-boat's magnetic signature on the earth's magnetic field. The US also brought the sonobuoy into the war. These operated using similar acoustic principles to the hydrophone in that they detected submarine noise and could be laid permanently or tactically in lines to provide acoustic warning of any submarine presence. MAD and sonobuoy systems were not deployed by the Royal Navy until after the war.

The primary ASW attack asdic set to emerge during the war was the Type 144. Development work started in May 1941 and twelve months later the first sets were in service. The design of the set was such that it could be modified to carry out different operational functions in a range of vessels being used in the ASW role but, most significantly, it was the key to the introduction of ahead-thrown weapons onto escort ships. Type 144 could cover an 80° sector either side of the bow and a sector search facility could be set automatically to step through either 5° or 2.5°. The range recorder was fitted with an optical cursor which could be lined up with returning echoes to measure the target range rate and this information was used to vary the paper speed on which the bearing was being manually recorded using four styli. The bearing recorder was mechanically more complex than the range recorder and used analogue computation to calculate the relative speed of the target across the line of sight. This allowed a trainable mortar to be aimed at the target's future position or the attacking ship to be steered over the target's future position for a conventional depth charge attack.

The Type 144 still suffered from loss of target if the submarine went deep or in the final phases of a depth charge attack. In 1943, this problem was mitigated by the introduction of the Q attachment which was a small transducer fitted below the main transducer. Q transmissions were synchronised to those of the Type 144 and used a much higher frequency of 38.5kHz than the main set. The transducer was mounted at 15° to the horizontal and had a beam width in azimuth of 3°. However, the vertical beam width gave coverage down to 60°, an increase of some 50° over the Type 144, and this allowed contact with the target down to depths of 700 feet at ranges of under 1500 yards. The

Hedgehog antisubmarine
mortar firing *c.*1944.
(Museum of Naval
Firepower)

Q attachment measured a slant range but still did not address the problem of calculating submarine depth. This was eventually resolved by the development of Asdic Type 147.

In mid 1942, in conjunction with Type 144 installations, the Hedgehog mortar started to be fitted to convoy escorts. Hedgehog was configured to fire salvoes of up to 24 mortars in a circular pattern over the ship's bow. Each projectile was filled with 30lbs of TNT and fitted with a contact fuze. Hedgehog had been selected, not without controversy,[89] as the weapon for the Royal Navy's ASW capability and formed part of the world's first integrated ASW system. With the submarine's position, course and speed now available, the ship was steered towards a predicted target position and a salvo of mortars was launched over the bow. During the approach and firing sequence, the U-boat remained in asdic contact and depth was not required as actual mortar contact was needed to achieve detonation. While any failure to detonate allowed the asdic to continue to hold contact, the fact that the there was no psychological impact on the submarine's crew was, itself, thought by some to be a critical shortcoming in the total effectiveness of the system.

More than 100 ships were fitted with Hedgehog by the end of 1942 but operational results were poor despite the Admiralty's instructions to use Hedgehog in preference to depth charges, something which ships had not been doing because of problems with serviceability and drill operations. Doubts were also raised about the adequacy of the warhead size against the U-boat pressure hull and the fact that the system's usefulness was limited to a deliberate attack

Loch Class Squid antisubmarine mortar mounting. (Collingwood Museum)

scenario with no capability for an urgent attack, due to the need to line the ship up with the predicted target bearing.

In May 1943, prototype trials of Asdic Type 147 were successfully carried out in HMS *Ambuscade* and Type 147 was brought into service in September 1943. This set provided a depth-finding capability using a streamlined sword-shaped transducer that could be lowered in front of the main Type 144 transducer and angled down to track the submarine. At 50kHz, the operating frequency was much higher than the main set and made the path of the generated sound through the water less prone to bending through any temperature variation with depth. Measurement of the slant range of the returning echoes, coupled with the known depression angle of the sword transducer, enabled submarine depth to be computed. Type 147 asdic was fitted in new installation ships in combination with the Squid AS mortar system.

The Squid mortar was actually a derivative of one of the heavier mortar options discarded in 1941. It comprised two triple-barrel mountings, installed forward either on the focsle deck or one deck higher, and it was capable of firing six mortar bombs, each containing 200lbs of Minol, in two triangular patterns out to 300 yards. The mortar had a sinking speed which was twice that of the Hedgehog projectile and the fuze could be set to activate at any depth down to 900 feet. The mounting was trainable by +/-30° either side of the bow. This capability provided stabilisation against ship roll and enabled some offset from the bow of the firing angle. With the submarine future bearing and range now available from the Type 144 and depth information from the Type 147, all parameters for a successful anti-submarine attack were in place, albeit still limited to within a few degrees either side of the bow. A further refinement of the Squid fire control system was later added whereby, in the case of a surface target and no available asdic tracking data, the aim and firing points for the mortar salvo could be calculated using surface radar tracking data fed into the Type 144 to allow a surface attack with a weapon detonation depth set at 20 feet.

The Squid system made its first kill on 31 July 1944 when HMS *Loch Killin*, a new Loch Class frigate, went to the assistance of HMS *Starling* and sank U-333.[90] Despite arriving more or less

Quadruple 2-pounder pom-pom fitted to HMS *Palomares*, a merchant ship converted for anti-aircraft defence. (Naval Museum Portsmouth)

at the end of the U-boat war, the weapon was credited with 17 U-boat kills and an attack to kill ratio of 50 per cent.[91] In both single and twin mounting configurations, Squid proved to be a very successful ASW system and remained in Royal Navy service until 1977.

Surface Weapons

The Royal Navy's experience against the Luftwaffe at Dunkirk and during the failed Norwegian campaign made it clear that their ships were vulnerable to air attack. The close-range guns adopted pre-war for air defence, namely the QF 2lb pom-pom, the 0.5-inch Vickers machine gun and even the .303-inch Lewis machine gun had all proved inadequate in terms of stopping power and unable to cope with diving aircraft at high elevations.

The Swiss-manufactured 20mm Oerlikon had been identified as a possible replacement for the Vickers machine gun as early as 1937 but it was never formally evaluated, despite the case being put forward by Lord Louis Mountbatten. When Admiral Sir John Backhouse was appointed First Sea Lord in November 1938, Mountbatten was at last successful and an order was placed for 1500 guns. Only 109 guns were delivered before the French surrender in June 1940 and these were used for land-based anti-aircraft use. However, the Oerlikon Company did grant a licence for the production of the gun in the UK and by the end of 1940 the first British-made guns were being produced at Ruislip, London. The Royal Navy started to take delivery for ship fitting around March 1941.

A more effective alternative to the heavier calibre pom-pom was found in the 40mm Bofors, a Swedish single-barrel gun already in use by the British Army for anti-aircraft defence. After the withdrawal from Norway, some of the Army weapons were quickly adapted for use in Royal Navy ships. The early mountings were not stabilised and this made seagoing use difficult. However, the Dutch Navy had a stabilised mounting in service and brought the design to England when their remaining ships regrouped in the UK after the invasion of Holland. By 1942, British-built stabilised mountings were being sent to sea in single and twin barrel form. As the war progressed, fire control of the Bofors mounting evolved with a range finding radar Type 282 being fitted along with remote power control and a tachymetric prediction facility to calculate the aim-off required to allow for aircraft speed.

The Second World War saw the beginning of the end of the era of the battleship as its tactical role reduced and the threat from aircraft and submarines increased. Both of these platforms were able to exploit the weaknesses in the design of capital ships, which had largely been conceived when battles between grand fleets were still envisaged. With no equivalent opposition, the use of the major calibre gun was starting to be limited to shore bombardment, an operational task which was in itself confined to the early phases of establishing a beach head. In addition, capital ships laid down during the war to make up for early losses never made it off the blocks as the build time was too long and the cost prohibitive.

One development, which prolonged the useful life of both these guns and enhanced the capability of the smaller calibre weapons, was the proximity fuze. Previous fuzes had been of the direct action type – which required contact with the target to achieve detonation – or the fuze burning and time-mechanical variety which respectively used a timed burning fuze train or a clock mechanism to count down a predetermined time of flight before initiating shell detonation. The time set on such fuzes was calculated using the associated fire control computer and set by hand or machine just prior to loading the shell. Although the emerging gunnery radars had improved the quality of range and bearing inputs into the fire control solution, this had only served to expose other sources of error in the computation process, which later proved to have equally, if not greater, degrading effects on accuracy. Thus, while radar did much for anti-aircraft gunnery, system performance in this mode continued to disappoint.

40mm Bofors twin mounting fitted for escort carrier anti-aircraft defence. (Naval Museum Portsmouth)

Mechanically setting the fuze itself introduced some six seconds of delay into the prediction process and exacerbated all other sources of error not yet even factored into the anti-aircraft fire control computation; such factors included gun throw-off and variations in the target flight path away from the straight and level.

The proximity fuze went some way towards mitigating against some of the errors in anti-aircraft fire control prediction and, more importantly for the big gun, improved the effectiveness of high explosive shells by ensuring that they could be exploded above the target, ship- or shore-based personnel and cause maximum incapacitation.[92] The fuze unit was fitted in the nose of the shell and contained a small radar that transmitted electro-magnetic waves. When in the vicinity of a target, these waves were reflected back from the aircraft or any other solid surface and detected by the fuze. When mixed with the transmitter frequency, the reflected frequency produced a beat frequency due to the doppler change induced by the speed of the target. At a predetermined doppler signal level, the detection caused the arming process to commence and the shell was detonated. The proximity fuze, also known as the VT fuze,[93] proved very effective against airborne targets with steady flight paths such as the V1 and it played a significant part in dissuading Germany from continuing with its indiscriminate flying bomb attacks on London.

The fuze was also found effective against kamikaze attacks during the Pacific War, again due to the straight-line flight profile and low bearing rates experienced during such attacks. Although the Allies were initially reluctant to use the VT fuze in situations where a dud shell could allow the fuze design to fall into enemy hands, during the early stages of the battle to retake France it was eventually released for use in the shore bombardment of land forces where the high explosive detonations, just above ground level, proved to have a devastating effect on exposed personnel.

CHAPTER 5

ELECTRICAL ENGINEERING MANPOWER 1939–1945

Pre-War Shortfalls in Electrical Specialists

Since the introduction of the EA in 1912 there had been very little advancement opportunity for the EA above the rate of Chief EA. Although the first Warrant Electrician had been promoted in 1911, the numbers had been small and, in March 1939, there were only 50 shown in the Navy List and of those 45 were promoted during the build up to war in the period 1936–1938. Consequently, the numbers available for promotion to Commissioned Electrician were also restricted. In fact, only 19 Commissioned Electricians existed at the beginning of the Second World War with none shown as being promoted during the period 1929–1937. As yet no Electrical Lieutenants had been promoted from Commissioned Electrician.

So it was that, in 1939, the Royal Navy again found itself with a shortage of general electrical engineering skills at officer and warrant officer level. Although various executive officers had gained some specialist technical experience in the Gunnery, Signals, AS and Engineer Branches, the overarching experience of electrical maintenance lay in the hands of the beleaguered Torpedo Officer, who was designated as the ship's expert,[94] but he also had warfare duties to contend with.

In his electrical role, the Torpedo Officer had EAs and LTOs to fulfil both his ship-wide responsibilities and specialist branch responsibilities. Employed by the Torpedo Officer as a skilled electrical tradesman, the EA made up the ship's general electrical maintenance and repair capability giving support to the technical specialists in other departments as necessary. In his duties the EA was supported by the LTO(LP) who was capable of semi-skilled maintenance and repair work on some general electrical equipment.

Experienced technical specialists in WT, AS detection and gunnery equipment existed in the Executive Branch at the warrant and commissioned officer level but few had the in depth technical training to enable them to adjust to the maintenance of new technology and, in any case, their time was more than spoken for when maintenance tasks were added to their branch operational duties. It is also worth remembering that the branch technical specialist was only responsible for his equipment from the main power distribution system supply fuses. Given the increasing complexity of the ship's electrical system plus the problem of conversion machinery being added to power new equipment, the detailed understanding of the system operation under adverse damage control conditions was becoming vested in, arguably, too few experts. Once again the Admiralty Board had reason to call on the RNVR for technical help.

RNVR Electrical Officers

In order to fill the shortage of officers with electrical expertise, the RNVR electrical officer category was created with candidates being recruited from the civilian electrical engineering field. Generally, these officers were electrical engineering graduates who had been involved with technology being developed by industry before the war. In common with all RNVR officers, they were known a 'Wavy Navy' officers because of the zig zag gold lace stripe worn on their uniform, and they also wore dark green distinction cloth stripes, alongside their gold stripes, to denote that they were electrical specialists.

At sea, the RNVR electrical officers usually reported to a departmental user officer for the conduct of their duties. In addition to their practical expertise, they were given responsibility for managing the electrical mechanics who had been brought into the service as Hostilities Only personnel. Many RNVR electrical officers were eventually to take up regular commissions as the importance of having career technical specialists, both ashore and afloat, became fully recognised by the Royal Navy with the introduction of the new Electrical Branch in 1946.

RNVR Special Branch Officers

The rapid increase in radar technology deployed in the Fleet from early 1940 was of particular concern to the Admiralty as there was no technical officer corps capable of taking on the upkeep task at sea, unlike in the WT, gunnery, asdic and torpedo fields where specialist executive officers had undertaken some technical training relevant to their maintenance responsibilities. The Torpedo Officer, who in most cases had ultimate responsibility for providing electrical support to other departments, was himself not an electrical engineer and, given his operational load, was not in a position to take on support of what were, in many cases, new concepts in technology terms. In fact, it is pertinent to point out that such was the novelty and secrecy of radar there were no university courses available in the UK at the beginning of the war likely to generate any graduates with even a basic knowledge directly suitable for radar work.

Initially, physics, electrical engineering and mathematics graduates, as well as more mature engineers from academia and industry, provided the main sources of recruiting and one of the key requirements was an aptitude for the concept of radar. Once selected, candidates were recruited as either civil service officers into the military radar research and development establishments or commissioned officers in one of the three services. However, many of these recruits still had to learn on the job as universities had yet to set up radio physics courses. The Signal School started to run its own specialist courses for radar officers in 1940, providing the training programme before officers were sent off for shore or sea duty. Such was the scarcity of these recruits that, in 1939, a tri-Service agreement placed responsibility for all recruitment into the radar field with the Royal Navy under F. W. Brundrett who was Assistant Director Scientific Research and Experiment (Admiralty). Brundrett was tasked with the equitable sharing of recruits between the services and, particularly, ensuring that posts in research and development establishments were best filled to meet staff requirements for radar equipment.

"BUT IF YOU DONT EXPLAIN RADAR To HIM, JOHN, I'M AFRAID HE'LL PICK IT UP FROM THE BOYS IN THE STREET."

(Collingwood Museum)

When the RNVR was reformed in 1923, it included the formation of a Special Branch (RNVR(Sp)) which was to be used as a means of engaging any specialist support needed by the Royal Navy. However, to denote that these officers were 'specialists' they also wore an emerald green distinction stripe alongside the gold.[95] It was into this RNVR branch that RDF officers were recruited in 1939.

Electrical Artificers

The EA rate had been introduced in 1912 and still formed the core electrical engineering capability of the Royal Navy during the Second World War. After a craft apprenticeship, the EA went to *Vernon* as an EA 5th Class to learn about the maintenance, fault diagnosis and repair of a wide range of electrical equipment from heavy machinery to delicate instruments. The detailed equipment training in *Vernon* could also cover gyros and the mechanical engineering aspects of torpedo and mining installations if the EA in question was to be given the maintenance responsibility for such equipment on board ship.

Electrical Artificers (Radio)

Initially, maintaining RDF was very much the prerogative of the RNVR specialist officer but it soon became clear that he would need some form of back up to take on the daily maintenance of not only RDF but also the RCM, communications and asdic systems being brought into service. Because of the shortage of ratings with radio experience, the Electrical Artificer (Radio) (EA(R)) rate was introduced and these men had to be brought in as Direct Entry recruits. By the end of the war the Royal Navy's own Artificer Apprentice Scheme had been expanded to include both the EA(L) and EA(R) specialisations. These apprentices were categorised as either 'electrical' or 'radio electrical' during their time at *Caledonia* and before they proceeded to their specialist training courses at *Vernon*. The *Vernon* training time for the EA(L) was nine months but for the EA(R) it was eighteen months and this meant that the trainee throughput was insufficient to meet the requirement of the Fleet until very late into the war.

(Collingwood Museum)   **REPAIR OF TRANSFORMERS AT SEA**

Electrical Artificers (Ship Repair)

In 1944, with the war against the Japanese in its final stages, extra ship electrical repair facilities were required in the Leyte Gulf, the Sakashima Gunto group of islands in the Pacific and in the main base in Sydney, Australia. Accordingly, electrical tradesmen from both naval and private dockyards in the UK were conscripted and a new breed of EA appeared, suffix SR (Ship Repair). EA(SR)s were only liable for service in shore repair bases or depot ships such as HMS *Resource*, which provided repair support for HMS *Indefatigable* in the Leyte Gulf after she had been damaged by Kamikaze attacks during the invasion of Okinawa.

Electrical Mechanics

Throughout the war, electrical training resources were at a premium and, arguably, too centred in *Vernon* at the expense of other port-based training schools. Although a sufficient number of sailors were being trained as Seaman Torpedomen and the LTOs were being produced to meet

Workshop training on aerials at Rugby Technical College. (Collingwood Museum)

(Collingwood Magazine Archives)

the wartime demand,[96] the training of a skilled EA was lengthy and intensive and the required numbers could not be met by men coming through the existing apprentice scheme. In order to boost numbers, the only option was to bring in skilled recruits who already had some electrical engineering experience and to adapt that experience to the naval environment.

Unfortunately, this type of recruit was invariably employed in the reserved engineering occupations and special release authorisation for service in the military was required for each one. This made the recruiting pool very difficult to access but, in 1941, the Electrical Mechanic (EM) (Hostilities Only) rate was introduced. The wartime EM was categorised as an 'artisan' rate because it did not encompass all the skills of the artificer rate but was deemed to be more skilled than the torpedoman rate. While not fully conversant with service conventions and with a limited technical range, the EM provided much needed technical help to the EA at sea. Under this direct recruiting scheme, suitable rating conscripts with some technical background were invited to become EMs and, after initial basic naval training, they were sent to *Vernon*(R)[97] and given similar training to an EA 5th Class in electrics. Once qualified, the EM was rewarded with advancement to leading rate and given the privilege of wearing fore and aft rig uniform before being sent off to sea.[98]

Wireless Mechanics

The problem of recruiting suitable trained men from industry as EA(R)s and the associated eighteen-month naval specialist training time meant that there was a serious shortfall of ratings with radio expertise at the beginning of the war to back up the RNVR officer. This requirement was met by establishing another artisan rate, that of the Wireless Mechanic (Hostilities Only). As with his EM counterpart, the Wireless Mechanic was recruited from industry but he had also to demonstrate some aptitude and ability for what was a rapidly advancing field of technology. To do this, the aspiring recruit was expected to produce some evidence, however slight, of studying, maintaining, operating or having anything whatsoever to do with radio technology in any shape or form! If a modest connection could be demonstrated, then he was sent off to HMS *Valkyrie* on the Isle of Man, an establishment which had been a pre-war holiday camp but was now being used for radio electrical training. Once trained, the Wireless Mechanic had a career path similar to his counterpart with accelerated opportunities to reach petty officer status.

In 1942, the Wireless Mechanic rate was re-designated as a Radio Mechanic rate. In this rate, he could now specialise in radar or shore wireless technology with specialist qualification (SQ) suffices of (R) and (S) respectively. Another SQ course in small ship radio systems was also introduced to further categorise the Radio Mechanic(R) to a Radio Mechanic(W) and give him the flexibility to undertake radar and radio maintenance duties in both

WRNS HFDF Training at HMS
Mercury. (Collingwood Museum)

radar and signals departments at sea. By 1943, it was found that the breadth of craft skills required to support the ever increasing range of radio and radar equipments at sea was too great for one individual and separate specialist qualifications for radar, radio and shore wireless with SQs (R), (WR) and (WT) were introduced.

Such was the scarcity of craft skills in this technology field, the Women's Royal Naval Service (WRNS) also introduced Radio Mechanics with similar SQs for employment in shore-based maintenance duties. Both male and female training schemes followed the same pattern with recruits spending some twenty weeks at technical college for basic technical and craft training followed by four weeks at HMS *Scotia,* on the Clyde, and ten weeks at the Signals School, HMS *Mercury,* for in service equipment training.

Rating Wiremen

In addition to the need for advanced technology specialists, the challenges of the Second World War again exposed shortcomings in the Royal Navy's basic electrical installation expertise and, once again as in the First World War, the Admiralty had to introduce the rating wireman. Although not an electrical engineer, such a tradesman invariably had experience in wiring installations for buildings, industrial machinery and motor vehicles. Although he had no direct experience of Royal Navy ships, the rating wireman was employed to great advantage, mainly in small ships, minesweepers and repair bases. The main tasks of the wireman not only involved the rewiring of ships after battle damage but also the support of ad hoc installation programmes for new equipment. Although employed for their electrical skills, wiremen were still categorised as members of the Torpedo Branch and still had a primary allegiance to the associated warfare duties of that branch. This was in spite of the fact that their work was heavily biased towards general electrical systems rather than underwater weapon technology. Even though their Service engagement was intended for hostilities only, many rating wiremen were advanced to leading rate and some went on to achieve petty officer status.

Headquarters Electrical Responsibilities

During the war the Admiralty had started to adopt a policy of separating the naval staff function[99] from the supporting materiel providers and the Fleet, including the training establishments. This gradually led to the formation of naval staff divisions to identify the requirements and materiel departments to come up with equipment solutions. As far as matters electrical were concerned, the division of responsibility for materiel in the Royal Navy can be summarised as follows.

Within Admiralty headquarters, the Director of Electrical Engineering (DEE) was responsible for the design and production of all electrical machinery except apparatus making use of electromagnetic waves. DEE was also responsible through the dockyards for the installation of all electrical machinery and apparatus in HM Ships and assisted with production and overseeing of electrical materiel dealt with by other departments. Its Director continued to be the Electrical Engineering Advisor to the Admiralty Board.

The Director of Radio Equipment (DRE) was responsible for the research, development and production of all communications and radar equipment. This included direction finding and radio navigation aids, where the responsibility had been transferred from the Director of Signals Department when it had become a division of the naval staff. That change meant that the Signal Branch no longer had the materiel responsibilities for radar which it had held for the majority of the wartime period.

The DNO held responsibility for the design of fire control apparatus and for the general layout and installation of fire control and associated communications systems. It was probably under this remit that the department was given responsibility for underwater fire control apparatus, including the Hedgehog ASW system.

Ship Departmental Electrical Responsibilities

At sea, the Torpedo Department was responsible for the maintenance of all high and low power electrical apparatus on board with the exception of high power equipment provided for the sole use of the Engineer Department. The Torpedo Department was also responsible for maintenance of torpedoes, all forms of countermeasure against enemy underwater weapons, controlled mining and guard loops. As the electrical expert, the Torpedo Officer held ship-wide responsibility for the overall electrical efficiency of the ship and was obligated to advise any other departmental officer of the need to make good general electrical defects, such as earthing defects, as well as provide electrical support when requested. In his role of electrical specialist, the Torpedo Officer was tasked with advising DEE on the design suitability of the electrical apparatus installed in ships.

The AS Department was responsible for operating and maintaining submarine detection apparatus, echo sounding, and indicator loops. The maintenance of asdic was carried out by senior detection ratings who reported to the AS officer.

The WT Department operated all radio and HFDF equipments and, where headed by an officer qualified as a signals specialist, that officer was also responsible for the maintenance of all WT equipment. The WT Officer was empowered to seek technical assistance from the Torpedo Officer in making good electrical defects. Before the arrival of the RDF officer, the WT officer was also responsible for RDF operators, who were usually wireless telegraphists or Hostilities Only RDF ratings.

In the early stages of the war with the rapid fitting of radar installations, the actual maintenance of the equipment was undertaken by specialist RNVR RDF officers. These officers were given direct reporting access to the Captain due to the secret nature of their tasking. As the radar maintenance load increased rapidly from 1940, more RNVR RDF officers[100] started to join ships to support the torpedo officer. Unfortunately, the numbers were soon insufficient to keep up with routine maintenance tasks on board and, from 1941, Wireless Mechanics who had been recruited directly from industry started to appear in greater numbers for the day-to-day radar support. However, in the event of problems, particularly in the small ship groups, RDF Officers found themselves being transferred around ships for repair tasks and, on occasions, operating duties because of the novelty of much of the equipment. In July 1943, the term RDF was dropped by the Royal Navy and the responsible personnel were thereafter designated as radar officers and ratings.

(Collingwood Museum)

YOU WIV YER 'ANDS IN YER BECKETS LIKE A BLEEDIN' RADAR OFFICER! — GIT CRACKING!

The engineering department had a small staff of ERAs and stokers charged with the maintenance of electrical apparatus installed in the ship's machinery spaces. Apart from some general electrical training at Keyham for officers and *Caledonia* for artificers, both officers and ratings were trained in ship high-power electrics at the Torpedo School in *Vernon*.

Curiously, the situation had evolved differently in submarines where the responsibility for electrics had fallen within the remit of the First Lieutenant rather than that of the Engineer Officer. This was despite the fact that the latter had been part of the submarine scene from the start and might logically have been expected to be electrically competent given the electrically powered nature of the boat's propulsion system. The T Class was one of the largest classes of submarines in the Fleet and the wartime build programme was prolific. The more generous size of the hull allowed some of the significantly more complex electronics to be fitted and this led to the addition of RNVR electrical officers at sea in submarines for the first time. Having taken his place in the complement, it was assessed that the new electrical officer was technically underemployed and, to earn his keep, he was expected to share fully in watchkeeping duties with the executive officers, both on the surface and when submerged.

Electrical Training Pre-1946

Until 1946, the electrical training of departmental officers, including the those of the Engineer Branch, was usually confined to generalist courses made available for those officers who could expect to have some responsibility for electrical equipment at sea through the employment of EAs, electrically trained engineer ratings or torpedomen. This training was conducted in *Vernon* and at other specialist schools around the country. After 1939, specialist RNVR officers were expected to have a good professional electrical engineering knowledge on entry and were given the minimum of training on actual equipment before being sent out to fend for themselves.

Prior to 1946, training for the EA started with four years of apprentice training in company with all new entry ERA apprentices and, from 1918, OA apprentices at *Caledonia*, near Rosyth. During these apprenticeships, *Caledonia* concentrated on producing skilled tradesmen who were competent in fitting and turning and many other skills of hand, but little time was spent on electrical training. After *Caledonia*, the EA apprentices were drafted to *Vernon* as EAs 5th Class for approximately ten months of general electrical training and specialist training on equipment installed in Fleet.

The regular service Artificer Apprentice Scheme had been expanded during the war to meet the additional requirement for both EAs and EA(R)s. Some of the general electrical content of the course was common to both specialisations and this training was carried out at *Vernon* with the EA(R) then going on to complete radio electrical training courses at HMS *Collingwood*. This led to a total training time for artificers ranging from four years ten months for the EA to five years six months for the EA(R) in return for a commitment to a twelve-year engagement. In effect, when corrected for 'man's time',[101] a fifteen-year-old EA(R) apprentice would only spend nine years of his twelve years in the service as a trained man. Not only did this represent a huge investment in technical training but it also meant that by the time EA(R)s had completely passed through the basic training pipeline and actually joined the Fleet, the war had virtually finished.

Specialist Training Sites

During hostilities, electrical equipment training was not confined to *Vernon* but was dispersed throughout the country in establishments providing specialist facilities, often very limited, for any branch that had some electrical equipment operation or maintenance support responsibility. The primary reason for such a wide dispersal was the need to escape the bombing threat to the main training establishments at Devonport, Portsmouth and Chatham. However, another reason was due to the rapid advance in radio and radar systems technology achieved during the war. New

Fleet equipment was being fitted at an astonishing rate and this had generated requirements for new training facilities that needed to be established quickly and somewhere which was suitable as a temporary training establishment. These sites included the Admiralty research establishments, the more remote naval establishments and even manufacturers' premises, as industry was often the only source of the necessary technical knowledge and equipment.

One such site was *Collingwood* which had been constructed as a new entry training establishment for 'Hostilities Only' ratings of the Seaman Branch. The site comprised 197 acres of farming land that had been acquired in 1939 under a compulsory purchase order for £7,290. The area was 'the finest bit of corn land in the south of England' according to the landowner but it was better known by the locals as the best snipe marsh in the country. Proof of the latter was everywhere as the wooden accommodation huts had to be mounted on concrete plinths to prevent the ingress of water and vermin. Legend has it that each single-storey building required a concrete raft four times the size of its floor area to keep it stable and that sea boots were essential footwear for anyone who strayed from the connecting pathways.

The establishment had been opened on 10 January 1940 under the command of Commodore Cyril Sedgewick, a retired Rear Admiral brought back into the service for hostilities but with the rank of commodore. Seamanship training started a week later. Despite the marshland problems, the site had been well planned with buildings spread out to reduce the number of casualties in the event of an air raid and arranged to form four identical self-contained training sections, each with

Aerial view of the *Collingwood* training establishment, 1940. (Collingwood Photographic Department)

The first Captain of HMS *Collingwood*, Commodore Cyril Sedgwick, hands over command in 1944. (Collingwood Museum)

VHF radio workshop instruction at HMS *Scotia*. (Collingwood Magazine Archives)

a set of HQ buildings. Each section had its own administration office, duty officers' cabins, petty officers' accommodation and a drill shed. It was designed to accommodate 2,500 men and had a dining hall capable of feeding 600 men at a time.

In June 1940, WT training was transferred from HMS *St Vincent* into *Collingwood* and ratings were drafted to a radio school under the command of their own captain. A radar school was added in 1944 under the command of a Captain (Radar Training). Such was the secretive nature of radar technology at that time that, apart from the school Captain, all the radar school officers were RNVR and therefore not allowed to use the wardroom. This meant them having their own self-contained, top security annex situated by the radar school parade ground. The establishment of the radio and radar schools required a number of radio and radar systems to be installed for both operator and maintainer training and, eventually, EA(R) apprentice training.

Another site was HMS *Marlborough*, the name given to the wartime training facility set up at Eastbourne College after the college students had been evacuated to escape the bombing. *Marlborough* was commissioned on 24 September 1942 under the command of Acting Captain (L) G.E.A. Jackson and was tasked to carry out electrical, torpedo and mine training when it was moved away from *Vernon*.

At the cessation of hostilities in 1945, there was a requirement to return these commandeered properties back to civilian use. Coupled with this, the parlous state of the British economy dictated an urgent need to rationalise training back into the main Royal Navy port areas as soon as possible. The scale of the task can be imagined in that there were some sixteen electrical training establishments with naval ship designators and many other minor facilities confined to single equipment training. A list of establishments is provided as an appendix with the caveat that some have undoubtedly been missed having been forgotten over time.

Thoughts of the Time

Vice Admiral Sir Philip Watson writes of the war-time period:

> During the hostilities, at the same time as the inadequacies of the ship's main electrical systems in resistance to damage were being revealed, rapid advances in electrical technology, which were taking place at an increasing speed accelerated by the needs of war, were enabling significant improvements to the fighting capabilities of warships and aircraft to be introduced. Most notable of these were the development and deployment of radar, voice communications systems at very high radio frequencies, integrated wireless communications systems and the application of electronics in the fields of gunnery and torpedo fire control, sonar systems and within the weapons themselves. Electrical systems to counter new threats, such as the degaussing system to counter the magnetic mine and the acoustic mine countermeasures, had to be developed quickly and introduced into service with all speed.
>
> As an expedient, young professional electrical engineers were recruited into the Royal Naval Volunteer Reserve as electrical officers to support the Torpedo, Signals and AS Officers at sea and from the shore. They were required to take direct responsibility in the fields of their technology which were entirely novel to the Royal Navy (such as radar). These officers were supported by recruiting electrical technicians and electrical and radio mechanics as 'hostilities only ratings'. Development in all these fields was rapid, and it soon became clear to the Board of Admiralty and others that the need for such expertise would be a permanent requirement and plans must be made to continue their activities into the future and beyond the cessation of hostilities.

With experience of the early war years, it became obvious that the projected development and fitting programmes for electrical and electronic systems in the Fleet would soon overwhelm the current electrical maintenance teams. In 1943, this caused the Controller of the Navy to issue a memorandum to the Admiralty Board that proposed the formation of a specialist Electrical Branch.

CHAPTER 6

TIME FOR CHANGE

The Admiralty Board Directive

The Controller's 1943 memorandum to the Admiralty Board proposed that two new branches should be formed, an Electrical Branch and an Underwater Warfare Branch. The memorandum stated that 'any reorganisation should improve efficiency in technical matters and should be economical of personnel'. It also proposed that a Flag Officer should be appointed to draft detailed proposals for implementing the scheme.

The memorandum was discussed by the Admiralty and Admiralty Board Minute 3886 approved, in principle, the formation of the two specialist branches as proposed. In broad terms, it was expected that the Electrical Branch would take over responsibility for electrical work from the Torpedo Branch and provide skilled personnel for the maintenance of electrical apparatus including radio, radar and underwater detection equipment. The Underwater Warfare Branch would then take over the Torpedo Branch's responsibility for all matters connected with torpedoes, anti-submarine weapons, mines and their countermeasures as well as being responsible for ASW, including the operation of asdic equipment and the development of tactics.

Rear Admiral, later Vice Admiral, H.D. Phillips was appointed to lead the study on 9 August 1943 and given six months to formulate proposals which would meet the requirements outlined in Minute 3886. The terms of reference of the study were:

1. To go into details of the proposed Underwater Warfare Branch.
2. To state the functions of the new Electrical Branch.
3. To report to what extent it is desirable to relieve Executive Officers of the responsibility for material and to recommend an organisation to take over these duties.

The Phillips Committee Report

On 15 March 1944, Admiral Phillips and his assistant, Lieutenant Commander Kilroy, submitted their report to the Board with the following summary recommendations:

1. The formation of an Electrical Branch to be responsible for all the duties now undertaken by the Director of Electrical Engineering, for the electrical work for which the Torpedo Branch are now responsible and for research, design, development, production and maintenance of radar, wireless and fire control apparatus, including the mechanical proportions of the latter which are at present the responsibility of the Gunnery Branch.

2. The amalgamation of the Torpedo and Anti-submarine Branches to form a Torpedo and Anti-submarine Branch. This Branch to be responsible for all matters connected with torpedoes and mines, countermeasures thereof, anti-submarine warfare, miscellaneous underwater weapons, devices and demolitions. The responsibility was to include the training and administration of the Radar Plot Ratings and taking charge of the action information centres of HM Ships.

3. The formation of an Ordnance Branch to be responsible for all (technical) matters connected with guns, gun mountings, directors, rangefinders, shells, cartridges, fuses, propellants, ballistics, high explosives and other gunnery equipments.

4. The concept of a Technical Officer in charge of all technical work on board ship with Specialist Officers reporting to him should not be implemented.

The fourth recommendation was made because at the time some senior engineer officers held the view that there should be only one technical branch with specialist engineers within that branch. Admiral Phillips rejected this argument on the basis that it required a breadth of training which was unrealistic at the time and that breadth was likely to increase in the future.

Vice Admiral H.D. Phillips. (Collingwood Museum)

Electrical Branch Proposals

Admiral Phillips recommended the formation of an Electrical Branch, the officers of which would have the same rank, status and pay as the officers of the Engineer Branch and carry the suffix letter 'L' after their respective ranks. Probably in anticipation of the perennial question, it was also stated that electrical officers would not be eligible for command at sea.

It was proposed that the Headquarters organisation for the Electrical Branch should take over all of the materiel duties of DEE and DRE. With this organisation in place, the control of the development of electrical, radar and fire control equipment was to be assumed by electrical officers with the supporting work being undertaken in the research establishments also under the control of an electrical officer. With regard to the existing civilian engineers working in these headquarters areas, it was recommended that they should be offered the opportunity of transferring into the Electrical Branch.

New entry electrical officers should enter the service as electrical cadets under the same conditions of service as executive, accountant and engineer cadets. They should serve one year as midshipmen and be employed in the same manner as Executive Branch midshipmen but with more emphasis on their technical training. After their time at sea as midshipmen, they were to proceed for further technical training at a Royal Navy establishment before then going to university to read for an electrical engineering degree. It was also recommended that junior electrical officers should expect to keep watches on deck in harbour.

The Electrical Officer should be supported by skilled EAs and a newly introduced category to be known as Electricians. Electricians would be 'more skilled than the present seaman ratings [torpedomen] by virtue of the fact that they would be devoting the whole of their time to the subject [electrical equipment].'

The Electrical Branch officers and ratings should be drawn from suitably skilled personnel serving in the Torpedo, Gunnery, Signal and Engineer Branches as well as the Fleet Air Arm. In addition, selected warrant radio telegraphists and RNVR radar officers should be allowed to transfer to the Electrical Branch and, subject to cross training, Air Artificer(L)s would be given the option to transfer to EA.

The Committee recognised the growing importance of radio and radar technology and that it would require a unique quality of specialist skill. Therefore, Phillips recommended that a new 'radio electrical' category should be established, alongside the existing 'electrical' category, within the proposed Electrical Branch. The radio electrical specialists would comprise the existing EA(R) and a new radio electrician rate and these would be the specialists in the radio and radar technology fields. While future EA(R)s would be recruited through the Artificer Apprentice Training Scheme, it was mooted that radio electricians could be recruited from artisans already trained in industry, the idea being that radio electricians would be skilled tradesmen, similar to the radio mechanics recruited for war service, but with more industry experience and naval training in this new field of technology.

On board ship, it was proposed that the Electrical Department should be formed to include the ship's existing radar equipment department. Apart from taking on responsibility for ship electrical, radar, communications and AIO equipment, Phillips also recommended that the maintenance of all Fleet Air Arm aircraft electrical installations should become the responsibility of the Electrical Branch. In order to meet the training requirements, electrical training schools should be established in all home ports.

Torpedo Branch and Anti-submarine Branch Proposals

Admiral Philips recommended that the Torpedo Branch and the AS Branch should amalgamate as soon as possible to form the Torpedo and Anti-submarine (TAS) Branch with the current junior officers of both source branches being given the opportunity to specialise in a dual capacity. TAS officers were also to have the opportunity to command MTB flotillas, destroyers and other vessels in which the AS role was predominant. In larger ships, it was recommended that the TAS Officer be in charge of AIO with an additional role of captain's staff officer on capital ships. Two rather novel recommendations were that TAS officers should be overborne in peacetime to allow for rapid expansion in wartime and that they should learn to fly.

The new TAS Branch was to retain responsibility for the technical aspects of its equipment, including torpedo tube design, and that ASW training and research should be carried out at *Vernon*, *Defiance* and the AS establishment at Portland when hostilities had ceased. The torpedo schools at *Vernon*, *Defiance* and Chatham should be equipped to meet the new training requirements for torpedo and anti-submarine warfare and action information organisation. The AIO training of radar plot ratings should also be administered by the new TAS Branch with technical radar training being undertaken in the home port electrical schools and the associated operational training being undertaken in the torpedo, gunnery and aircraft direction schools.

In anticipation of the major task of post-war mine and demolition clearance, it was recommended that a specialist section under a designated head should be established within the TAS Branch to undertake the recovery and disposal of mines and bombs.

Phillips also recommended the establishment of a non-executive, Torpedo Mechanical (TM) Sub Branch within the TAS Branch to fill posts in the torpedo and mining specialist area where it was felt that greater engineering knowledge would be required than that generally held by the warfare orientated TAS officer. It was envisaged that some TAS officers would be given engineering training to become qualified as TM specialists. In order to sustain the TM Sub Branch, a torpedo artificer category was to be introduced for torpedo technical support and be recruited under the existing artificer apprentice scheme. Arrangements were also to be put in place to provide a promotion route for torpedo artificers to reach commissioned rank. A number of naval staff, torpedo range and torpedo depot posts were to be filled by TM officers, preferably ex-torpedo artificers in the more technical posts.

One final recommendation by Phillips was that existing Torpedo Branch officers and ratings should be allowed to transfer to the new Electrical Branch.

Ordnance Branch Proposals

Phillips endorsed the formation of an Ordnance Branch with duties entailing responsibility for all matters connected with guns, gun mountings, directors, range finders, ammunition, fuzes and gunnery equipment for the Fleet and Fleet Air Arm. These duties were to include tasks currently undertaken by the Engineer Branch at sea on gun mountings but not the TAS Branch maintenance tasks on torpedo launchers which were to be in the remit of the new torpedo artificer. In addition, the branch should be responsible for the inspection of guns and ammunition, relieving the Chief Inspector of Naval Ordnance (CINO) of these duties and placing ordnance officers in posts within the CINO organisation. Recruitment of CINO staff should cease forthwith and the inspection of explosives carried out by the TAS Branch was to be undertaken by the Ordnance Branch.

In a similar manner to the new electrical officer cadets, ordnance officer cadets were to be recruited and trained under the same terms and conditions as the other branch cadets. Given the existing Engineer Branch responsibility for gunnery maintenance, suitably experienced engineer officers were also to be allowed to transfer to the Ordnance Branch.

The OA should be transferred from the Gunnery Branch into the new Ordnance Branch and a new rating of armourer should be introduced to replace the existing Quarters Officers (QOs) in ships. The rates of Air Artificer(Ordnance), Air Mechanic(Ordnance) and Air Artificer(Electrical & Ordnance) should be abolished and the associated duties should be undertaken by the Ordnance Branch. Leading Seamen(QO) and Able Seamen(QO) in the Gunnery Branch, who had experience of technical duties, should be allowed to transfer to the Ordnance Branch as semi-skilled maintainers. However, the QO rate should be retained by the Gunnery Branch but with their scope of employment being non-technical, quarters duties only.

Signal Branch Proposals

The Signal Branch responsibility for all radio apparatus maintenance should be transferred to the Electrical Branch, including the maintenance and development of wireless telegraphy and radar. Air signal officers and warrant observers were similarly to be relieved of their responsibility for equipment maintenance. Signal officers would retain responsibility for coordinating the use of all forms of electromagnetic transmissions at sea.

Selected warrant officer telegraphists should be given the opportunity to transfer to warrant radio electrician and RNVR radar officers should allowed to re-categorise as electrical officers after suitable cross training.

Engineer Branch Proposals

The main recommendation was that the new Ordnance Branch should take over gun mounting maintenance duties from the Engineer Branch and that a number of engineer officers with ordnance equipment experience should be allowed to transfer to the Ordnance Branch. Posts currently occupied by engineer officers in the Directorate of Armament Supply (DAS) should be filled by ordnance officers. Following the introduction of the TM specialist, the Engineer Branch responsibility for torpedo tube design should be transferred passed over to the new TM Sub Branch.

The Engineer-in-Chief strongly supported the idea that all technical branches should be under the supervision of one technical officer on board ship but Admiral Phillips eventually concluded that this was not appropriate.

The Admiralty Board Response

At an Admiralty Board meeting on 24 March 1944, a Committee was set up to scrutinise the proposals made by Rear Admiral Phillips. Within the scrutiny process, the Committee interviewed many authorities who would be affected by the Phillips' proposals and considered other options to the Phillips' recommendations.

One of the options examined was submitted by Admiral Sir Percy Noble, who was Head of the British Admiralty Delegation in Washington. Admiral Noble provided details of the US Navy's 'line officer' concept under which all officers received a limited common engineering training and were liable to serve in any departmental function aboard ship. Shore-based engineering posts were filled by selection from line officer volunteers, who were transferred to an Engineer Officer Duties Only Branch and given further technical training prior to taking up their engineering posts. Only in exceptional circumstances did these officers subsequently go to sea. However, the Committee felt that the US Navy structure would not be suitable as there was a 'fundamental difference of national character and outlook' and such a system placed 'a greater dependence upon subordinates than that to which our own Service is accustomed'.

The Committee felt, as a matter of principle, that maintenance support was best undertaken by those who used the equipment. While this was consistent with Phillips' proposal for a TM Sub Branch within the TAS Branch, it was not consistent with the introduction of a separate Ordnance Branch, or indeed the Electrical Branch. Furthermore, the Committee view was advocated alongside their somewhat contrary statement on the matter which said that 'Human limitations being what they are, it is in some fields necessary for efficiency that the responsibility for maintenance should be divorced from responsibility for use.' It was with this rather vague reconciliation of principle that the Committee then endorsed the institution of the Electrical Branch as proposed by Phillips and reported their conclusions on 15 September 1944. These conclusions were:

1. The institution of an Electrical Branch that would take on responsibility for the maintenance of all electrical apparatus and be accountable for that maintenance to the User Officer. The exception to this was that User Branches should retain responsibility for the maintenance of Armament Stores. With regard to Headquarters posts, while it was to be a long term objective that electrical officers should ultimately be capable of absorbing the functions of DEE, it was only to happen if it should become a requirement in the future.

2. The institution of a TAS Branch which was to be formed by the amalgamation of the present Torpedo and Anti submarine Branches. The TAS Branch was to include a maintenance sub branch that was to be designated as the TM Sub Branch.

3. The development of the existing Ordnance Sub Branch, within the framework of the Gunnery Branch, in order to provide greater assistance for the Gunnery Officer in the maintenance of all surface weapons [consistent with the endorsement of the Torpedo Mechanical Branch]. This assistance was to include the mechanical aspects of fire control systems.

The Committee did not support Phillips' proposals to increase the existing responsibilities for design, research and experimental work and the new branches were only to take over the existing responsibilities that Royal Navy personnel had for supporting civilian technical departments. The Royal Naval Scientific Service was to continue working with the new branches as it had done so in the past, thus avoiding any duplication of research and development facilities.

Similarly, the proposed changes to civilian arrangements for the ordnance inspection and armament supply were also rejected in order to retain these two functions in roles that were independent of both the users and the maintainers. However, Royal Navy posts within these organisations were to continue to be filled by appropriate personnel from the new branches.

The transfer of radar plot ratings and responsibility for ship's AIO to the new TAS Branch also failed to gain approval and these responsibilities remained with the Navigation Branch. Those radar ratings responsible for operating fire control radars were to be transferred to the Gunnery Branch. However, the maintenance aspects of the associated radar equipment were to be transferred to the Electrical Branch.

With regard to manning the new branches, the Committee had recognised the need for a semi-skilled technical rate but rejected Phillips' idea of recruiting an artisan rate as they felt it limited the prospects of advancement for junior ratings. Instead they recommended that the mechanician rate, which had existed in the Engineer Branch since 1905,[102] should be adopted by the new technical branches and used as a means of providing advancement opportunities for selected semi-skilled ratings.

The Fleet Air Arm was permitted to retain responsibility for the electrics and ordnance matters which affected air worthiness but the remaining technical responsibility was to be passed to the Electrical and Ordnance Branches.

On one final matter, not apparently addressed by Phillips, the Committee recommended that the Electrical Branch should take over responsibility for electrical machinery from the Engineer Branch.

The Resistance to Change

The Admiralty Board started to make known the outcomes of its deliberations in November 1944. Unsurprisingly, the promulgation of radical changes drew a wide range of criticism from both Government and Headquarters departments. Although at that stage of the war the cost of any organisational change was always going to be challenged robustly, much of the hostile reaction came from those who would clearly be the losers under the new structure due to reduced power or loss of professional standing.

Typical of the communications between the various departments is the following letter, dated 27 November 1944, sent from the Treasury Chambers to the Under Secretary of Naval Personnel at the Admiralty.

We had some talk on the telephone about your official letter of the 22nd of November 1944, regarding the proposed formation of an Electrical Branch in the Royal Navy, primarily with a view to taking over the electrical work on board ship now allocated to torpedo and other specialist officers. I appreciate that in modern warships there is a considerable field of activity for electrical specialists and I also appreciate the desire to retain the services of temporary officers who have shown specialist ability in this direction. Nevertheless, further consideration of this question in the Treasury has failed to convince us of the absolute necessity of embarking at this juncture on the rather formidable venture of setting up an entirely new Electrical Branch. I have not concealed from you our apprehension that the proposed creation of a separate branch with its own career and rates of pay will, in itself, involve us in substantial additional costs as compared with the present system, whereby the officers carrying out the work on board ship are merely in receipt of specialist allowances in respect of their particular qualifications. We are now even more alarmed at the prospect of the new branch taking over the duties on shore, including those at Headquarters, now performed by the Director of Electrical Engineering and his civilian staff. We would require more evidence than has so far been offered to convince us that the gain in efficiency thought to be secured by such a step outweighed the serious loss on economy which we fear. Our position therefore is that while we are quite ready to believe that more thorough training in electronics is required and perhaps also greater inducements to officers to specialise in this subject, we would very much prefer that the general basis of the present system should be continued, especially as regard work on shore. We hope therefore that the Board of the Admiralty may feel able to modify their present view as to the right solution of the present issue. I need hardly say that we should be very ready to cooperate with you in any measure designed to improve the electrical service on the lines we suggest …

The Staff side of the Whitley Council took up the debate on behalf of the Civil Service and expressed doubts as to whether electrical officers, despite the high professional qualifications that they would undoubtedly hold, would have sufficient continuity of experience to meet the requirements for design and development work. This would cause them to have limited credibility when dealing with industry at the risk of losing authority when trying to ensure that naval design requirements were being met.

The Engineer-in-Chief and his staff continued to question the economic effect on complements of employing a separate Electrical Branch, particularly to maintain and repair apparatus associated with a ship's Engineer Department. They advocated the continuation and expansion of electrical training then being given to Engineer Branch ratings, to cover all electrical equipment, with all associated work being put under the supervision of engineer officers suitable trained in electrical matters. The Committee did not support this view.

The Director of Anti-submarine Warfare (DASW) opposed the AS Branch amalgamation with the Torpedo Branch on the grounds that the new TAS Branch officers would be overloaded during wartime and that the AS Branch would be eliminated. These views were dismissed by the Committee as being without foundation.

The Director of Aircraft Maintenance and Repairs (DAMR) expressed a firm preference that all electrical equipment associated with aircraft be outside the orbit of the Electrical Branch as it was, after all, a General Service Branch with little understanding of the special nature of air engineering. They considered that the overwhelming case against the change was that the all-embracing tenet of 'air worthiness' was too important to be confused by potential issues arising due to any loss of Air Engineer Branch accountability for aircraft safety. This argument was accepted by the Committee.

DEE accepted that naval operational experience had a role to play in the equipment procurement process. However, while he considered that the Electrical Branch could take over the responsibilities of his department, he was sceptical that the level of engineering training they would receive would match that of the civilian officers currently in the department. The Committee agreed that naval training was at the time inadequate for such a change, accepted the DEE argument but retained the concept as an aspiration for the future.

DRE did not believe that it was possible to produce officers with sufficient qualifications such that they could competently direct the support activities for both electrical and radiating appliances. They also questioned the viability of recruiting ratings capable of reaching the standard of competence necessary for the future of the new Electrical Branch and postulated such a wide separation in training and application of personnel that a future division of the branch into two was inevitable, and in fact this did effectively occur with the later introduction of the radio electrical specialist category.

The Admiralty's Conclusion

Having listened to the reactions of the various interested parties, the Admiralty Board finally approved only the formation of the Electrical Branch and the TAS Branch as the result of the Phillips' report. The institution of the Electrical Branch was announced in AFO 517/45, dated 1 February 1945.

> In view of the increasing importance of electricity to the Navy and the high specialist qualifications required in personnel of all ranks carrying out electrical duties, Their Lordships have approved in principle the institution of a new branch of the Royal Navy to become known as the Electrical Branch. The Electrical Branch will be responsible for all electrical installations in the Fleet, except where special considerations require otherwise. Ultimately, it is expected that the Electrical Branch will be able to absorb the functions at present discharged by the Department of Electrical Engineering.

(Collingwood Magazine Archives)

Their Lordships hope that many officers and ratings on temporary service in the Navy who are carrying out electrical duties will be attracted to join this new branch of the Navy for permanent service. Selected personnel in existing branches of the Royal Navy will also be afforded opportunities for transfer if they so desire.

Conditions of service, qualifications for promotion and rates of pay in the Electrical Branch will be promulgated as soon as possible and volunteers will then be called for. This new branch will offer to officers and men alike not only a naval career in keeping with the importance of their work but also valuable opportunities for technical advancement. This preliminary announcement of Their Lordships approval to institute the new branch is made in order that officers and men may have opportunity to think the matter over in advance of the call for volunteers.

In February 1945, the Admiralty appointed Rear Admiral G.B. Middleton to chair a steering committee with the following remit:

To make recommendations to put into effect the decision of the Board to initiate an Electrical Branch of the Navy which will ultimately be able to absorb the present department of the Director of Electrical Engineering; also having regard to the report made by Rear Admiral Phillips on March 15th 1944 and the Board's conclusions thereon.

CHAPTER 7

THE MIDDLETON REPORT

Terms of Reference

The Middleton Steering Committee was tasked on 20 February 1945 to make recommendations to put into effect the decisions of the Board of Admiralty and establish an Electrical Branch of the Royal Navy. It was charged with responsibility for the technical and maintenance aspects of all electrical and radio installations and Fleet Air Arm electrical systems except where special considerations such as for air safety were paramount. This final caveat left an option open to the Middleton Committee to scrutinise again areas of specialist engineering matters, such as the airworthiness of naval aircraft which was still causing the Fleet Air Arm some concern. In addition to taking over specialist electrical support in the Fleet, the Steering Committee was tasked to structure the Branch such that it would be capable of taking over the Headquarters' duties of the Electrical Engineering Department in the future.

Vice Admiral Gervais Middleton. (Collingwood Museum)

The Middleton Principles

The Admiralty Board, having advocated that maintenance was best carried out by the user and, somewhat contrarily, endorsed the policy of a specialist maintainer by the establishment of the Electrical Branch, left the problem of deciding where the line between user maintenance and specialist maintenance should be drawn to Middleton. He concluded that the delineation of responsibility should be based on the level of technical training needed to support a maintenance task and this became a founding principle of his report. He decided that 'the point at which the maintenance burden could be said to be too heavy for a User Branch is reached when the technical training required by its personnel to enable them to perform maintenance duties is more than a small part of the total training.' This principle then led to a second principle that recognised a degree of similarity in the nature of any electrical maintenance problem and that there was always a basic level of technical training required to solve that type of problem. Thus any electrical specialist would need a significant core of basic technical training but, once completed, this would allow that expertise to be deployed to support any electrical equipment with only marginal increases in specific training. From this principle Middleton reasoned that 'The dispersion of effort entailed in the provision of separate groups of electrical technicians within User Branches is no longer sound in principle or practice' and to this he could have added 'cost effective'. He also acknowledged the emerging complexity of radio and radar technology and, perceptively, adopted a third principle which was that a conceptual understanding of radio technology was essential for maintainers in this field. It was with these principles in mind that the Committee set about the detailed implementation planning for the new maintainer branch and defining its relationship with the user branches.

Scope of the Electrical Branch

The responsibility of the Electrical Branch was to include the operation of electrical power supply systems and any electrically driven machinery in the engineering spaces, but in the case of the latter the electrical personnel were to be under the control of the Engineer Officer. The Branch was also to have maintenance responsibility for all electrical machinery currently being maintained by the Signal, Torpedo, AS, WT and Engineer Branches. There were to be air electrical and radio specialists within the branch but these specialists would be reporting to the appropriate authorities on matters of air worthiness.

Electrical officers were to be appointed onto the staffs of dockyards for ship support work, to shore establishments to carry out work on behalf of DEE and to shore radio stations for the maintenance of radio equipment belonging to DRE.

Branch Organisation and Structure

The Middleton Committee proposed that the Electrical Branch should be headed by a new headquarters department, designated the Naval Electrical Department (NLD), which in due course was to absorb DEE. NLD was also to take on the design and maintenance functions of DRE, the Department of Torpedo and Mining (DTM), the Department of Air Radio (DNAR) and DASW. Also, electrical officers were to be appointed to posts established in every other Admiralty department or research establishment that had an interface with electrical equipment.

Broadly, the structural aim for the Electrical Branch was to bring together the existing Royal Navy's electrical, radio and radar expertise under one umbrella branch organisation. Accordingly, officers and ratings were to be brought in as necessary from existing source branches in order to ensure that the appropriate technical support continued to be made available to meet the needs of the Fleet. The Branch was to be comprised of electrical officers, warrant officers, EAs and electricians with those men coming up through the lower deck being categorised as either electrical (L) or radio (R) specialists. A similar structure with a specialist suffix (Air) was recommended for the Fleet Air Arm.

Electrical officers were to be given the same status as offciers in other technical branches and to be distinguished by dark green distinction cloth between the gold lace stripes appropriate to their rank. They were to carry the suffix (L) after their rank, with the most junior rank being Sub Lieutenant (L). New entry electrical officers were to either to join the service as a cadet through Dartmouth, in a similar manner to existing branches, or through a university entry scheme similar to that already in existence for the Engineer Branch. The number of directly recruited officers would be supplemented by ratings who had been selected as being outstanding and having officer potential. New entry officers and, in due course, officers transferring into the Branch were all required to become qualified in the electrical, radio and radar aspects of their future duties. There was also to be a Branch List of electrical officers, for those promoted from the lower deck. These officers were to be given the rank of Commissioned Electrical Officer and they were to retain the suffix (L) or (R) depending on their specialisation on promotion. Warrant officers were also to retain their (L) or (R) specialist suffix.

The EA was to be categorised after a period of electrical training as either (L) or (R),[103] with the proviso that selection for the (R) category required the candidate to have demonstrated an aptitude for radio technology training.

The Electrician rate was to be based on the substantive rate of petty officer with advancement opportunities to Chief Electrician or Chief Radio Electrician depending on specialist category. The electricians were to be supported by junior rates, to be known as Electrician's Mates (EMs). The term 'electrician's mate' was adopted because the Middleton Committee felt that the title of electrician had artisan connotations and, if applied to junior ratings, would give the impression that Electrical Branch junior ratings were of a higher status than junior ratings in other branches. New entry junior ratings were to be recruited as 'provisional electrician's mates' and selected on aptitude during training to specialise in general electrics or radio electrics. The electrician's mate was to be given advancement opportunities to electrician or radio electrician based on experience, selection and further training.

The importance of having promotion and advancement opportunities throughout the branch structure was fully recognised and emphasised in the Middleton Report. As such, both artificers and electricians were to be given the opportunity for selection and promotion to both commissioned rank and warrant rank. In addition to standard advancement routes from electrician's mate to electrician, accelerated promotion opportunities were to be made available for exceptional leading electrician's mates to Sub Lieutenant (L).

Branch Manning

The manning of the Electrical Branch would initially have to be taken from existing Fleet resources, given that the equipment being transferred was already supported in some way. In the case of dedicated maintenance support officers and ratings, these personnel would be transferred into the new branch to meet the initial manning requirement. Electrical officers were to be recruited by the transfer of suitably skilled regular officers and warrant officers from the existing branches and RNVR electrical officers seconded to the Royal Navy for war service. Commissioned officer recruitment was also to be open to warrant electricians, although they were few in number, as they represented the most senior rank with regular service electrical expertise.

In order to meet the calculated officer manning requirement, the Middleton Committee determined that a complement of 220 officers was needed. Twenty commander posts were identified to form the initial planning team needed to set up the successful launch of the new branch, and Treasury authorisation was obtained in 1945 for the establishment of these posts.

The selected officers were to be given the suffix (L) after their substantive rank and they were to don the green cloth branch stripe on transfer. A number of other key posts were to be filled by selected officers with the remainder being filled by volunteers from other branches. An initial list of 71 officers, who had been selected for transfer to the Electrical Branch was promulgated in 1945. The seniority of the officers to be transferred was set at 1 January 1946, the scheduled launch date for the Branch.

The EA(L), as a rate, already existed in the service and these men had previously provided skilled electrical support to all branches whilst being under the management of either signals or torpedo officers. Hence, with the introduction of the Electrical Branch, those EA(L)s with responsibility for electrical systems being placed under the remit of the new branch were to be transferred from their existing line management officers and placed under the direction of the ship's Electrical Officer. The other source of electrical expertise available was the pool of Electrical Mechanics, who had been brought in for war service. Many had shown themselves capable of carrying out skilled electrical maintenance duties, and it was decided to give them the opportunity to transfer into the new branch as EA(L)s or as Electricians.

The support of the new radio and radar technology posed more of a problem. Very few regular officers had any in depth radio support experience and radio training had only been introduced into the Artificer Apprentice Scheme during the war, but the first EA(R)s were not yet available

(Collingwood
Magazine Archives)

LET'S SEND HOPKINS – HE'S ALWAYS FIDDLING WITH THE WARDROOM WIRELESS

to join the Fleet. The Signals Branch had a radio maintenance capability in its telegraphists which would help bridge the gap until the EA(R)s appeared, and these men needed to be encouraged to transfer to the Electrical Branch. The bigger problem was that at the time of the institution of the Electrical Branch, radar technology was being almost entirely supported by RNVR officers and radio mechanic ratings brought in for war service only. This meant that there was a need to appeal to the volunteer instincts among those identified as possible candidates for the Branch but who were anxious to return to civilian life. While enforced extensions to war service offered a short term solution, the better option was to make sure that the pay, conditions of service and career prospects were competitive with those in industry to encourage retention for a longer period.

Originally, most of these radio mechanics had some radio engineering training before entering the Royal Navy and many had accrued sufficient maintenance experience during the war to be considered worthy of artificer or electrician status, particularly in the conduct of their maintenance duties. As it was unlikely that all of these men would be persuaded into the new branch, it was expected that there would be shortages in the radio and radar field. In order to reduce this risk, suitable EA(L)s were also offered the opportunity to cross train in radio and radar engineering.

There was a significant element of maintenance support not covered by the existing EA(L)s, RNVR officers and mechanics and this was vested in those officers and ratings who also held some degree of executive or operator responsibility within their parent user branch. While Middleton recommended that these men should be allowed to transfer to the Electrical Branch, they had to be persuaded to volunteer and, if successful, this could create consequential manning problems for their source branches. It was anticipated that the latter was bound to be of some concern to those parent branches.

The Committee proposed that the senior rate electricians should to be selected from suitably experienced volunteers from the Torpedo Branch. In general, these were the chief petty officer and petty officer torpedo gunner's mates and the leading torpedo operators. These men would bring torpedo and low power equipment experience and responsibility with them from the ship's Torpedo Department over to the newly formed electrical department. While many of these ratings had virtually full time electrical and torpedo maintenance duties with no executive tasking to impact on the Torpedo Branch, there was still the unknown factor of branch loyalty.[104]

The majority of new electrician's mates were to be taken from the seaman torpedomen with electrical maintenance experience. This was more difficult, as many of these ratings would still be needed to carry out what would be ongoing quarters functions for their torpedo department.

After considering the Phillips Report, the Admiralty view had been that the Electrical Branch should adopt the 'mechanician' rate in keeping with the Engineer Branch. However, Middleton argued that

> ... the skilled rating could be produced by advancing semi-skilled ratings who show the necessary qualities which have been developed by experience and training in the service. This rating we

consider to be comparable with the Mechanician, making up in electrical skill what he lacks in skill of hand.' He also said that 'there is no room or employment for a petty officer (electrician) who remains only competent to undertake Grade 3 work.[105]

As such he was saying that a senior rate electrician should be capable of not only more substantive skills but also more non-substantive skills than his juniors.

Electrical Branch Training

The post-war situation of having electrical training dispersed about the country was to be resolved by the establishment of electrical schools in each of the home port areas of Portsmouth, Devonport and Chatham as well as at a selected naval air station. An additional school was to be provided at Rosyth but only if, in the future, it should be designated as a home port. HMS *Collingwood* was nominated to become the school for Portsmouth and the 'lead' electrical school for the service, with HMS *Defiance* becoming the Devonport Electrical School in addition to its role as a torpedo school. At the time there was no electrical school at Chatham and this needed to be established.

Ordnance and Torpedo Mechanical Branches

In 1946, the Middleton Steering Committee was further tasked to review the structure of the Ordnance and Torpedo Mechanical Branches. When studying the options for the two branches, Middleton quickly found himself with some very strong, divergent views within the two sub-committees set up to examine each branch. The detail of the arguments is too broad and insufficiently relevant to present here fully, but there was a clear shortage of widely supported recommendations in the report and several issues appeared to be at odds with the previous report by Middleton. This can, perhaps, be reconciled by the fact that Middleton was initially focusing on the institution of the Electrical Branch rather than technical branches as a whole. Thus, for example, the fact that the embryo TM Branch was proposing to have a torpedo artificer was not a matter that he would necessarily have factored into his initial report regarding the training and career structure of EAs. The Middleton Committee reported its findings on this subject to the Admiralty Board in April 1946 but the conclusions were not well received by the Engineer-in-Chief who wrote at the time:

> This very lengthy report, which is in effect three reports, one made by the Main Committee and two by a Sub-Committee, leaves the reader somewhat amazed. The difficulty is to see the wood for the trees and to understand exactly what is being proposed and why, and thereafter to be constructive.

In his report, Middleton recommended that a separate Ordnance Branch should be formed and staffed by suitably experienced officers transferred from the Executive Branch or the Engineer Branch. These officers were to wear a blue cloth branch stripe and be streamed to take up seagoing maintenance, research and development or ordnance inspection appointments. However, each type of appointment would require different experience and qualifications and this is where the affected Headquarters groups began to voice their objections. The rating structure was to comprise a limited number of ordnance artificers and a considerable number of seamen from the Gunnery Branch and Royal Marines with quarters officer experience.

Regarding the creation of a Torpedo Mechanical Branch as a proposed technical sub branch of the TAS Branch, Middleton recommended in his conclusions the retention of the torpedo artificer, which went against his earlier principles used to define the Electrical Branch and the conclusions of his sub-committee. The sub-committee had concluded that the torpedo artificer was unsustainable as a separate class of artificer and that a specialist four-month course would be sufficient for an EA to undertake the proposed maintenance duties on behalf of the TAS officer.[106] The benefit of this argument was that it would have removed the need to provide a career structure for the torpedo artificer, as the EA would continue to retain his promotion opportunities within the wider Electrical Branch. Without the torpedo artificer, there was no subsequent need to have the warrant TM rank as a promotion avenue. This resolved another problem unearthed by the sub-committee whereby they had considered that there would be few shore based posts suitable for warrant officers TM. This view was endorsed by a number of Headquarters organisations who had expressed concerns that the warrant officer would not be suitably qualified or experienced to fill the shore based posts which were currently being filled by torpedo officers or engineer officers.

Although Middleton overrode his sub-committee's conclusions in this area, he did put forward strong caveats that the sustainability of a small TM Sub Branch was doubtful and would leave the torpedo artificer with very limited career prospects compared with his Electrical Branch equivalents. He also acknowledged that there could be strong objections as to the financial acceptability of establishing such a small sub branch due to a perception that the TM higher posts were established on the basis of achieving a branch structure rather than fulfilling an operational need. In a final comment, Middleton made the statement that he still believed that, eventually, there would be the need for a single technical organisation but the time for such an organisation had not yet been reached.

The debate regarding the future of ordnance and torpedo support continued within the Headquarters departments until September 1946. Finally, it was decided that the Ordnance Branch should remain as part of the Gunnery Branch under the direction of an engineer officer who was to remain a member of the Engineer Branch. Skilled technical expertise was to continue to come from the OA, who would be supported by a newly created stream of semi-skilled ratings, known as Quarters Armourers, and these ratings were to be sourced from suitable seaman and Royal Marine personnel. The OA would be recruited under the artificer apprenticeship programme, along with the ERA, EA and EA(R) and carry out the final stages of his training with the ERAs at HMS *Caledonia*. The Fleet Air Arm ordnance organisation was to remain unchanged except for the introduction of an Air Armourer rate to support the OA who would have received training in aircraft gunnery material.

The case for a TM sub branch appears to have been rejected by the Admiralty as no evidence has been found of either its implementation or the introduction of the torpedo artificer. In fact, as recommended under the original proposals, the Torpedo and AS Branches were merged together on 10 October 1946 to form the TAS Branch.

The TAS Branch ratings were designated either as 'underwater control', with responsibility for sonar and anti-submarine tactical operations, or 'underwater weapons' and responsible for husbandry type maintenance of torpedoes and other underwater weapons. The TAS Officer was also supported by an EA who had been trained in the maintenance of the underwater weapons and the associated fire control systems.

An Assessment of the Middleton Report

Vice Admiral Sir Philip Watson gives his verdict on the work of the Middleton Committee as follows:

> Admiral Middleton rendered a very comprehensive report, delineating in great detail the scope and responsibilities of the (Electrical) Branch, its permanent structure, training requirements and precisely how it should be set up, drawing its resources from the current organisation with responsibilities in the field of electrical engineering.
>
> It was appreciated that, in order to maintain continuity of knowledge and experience, it was, in the first place, necessary to attract a number of the RNVR electrical officers. These would

be offered permanent commissions before they would otherwise be demobilised, in order to form the nucleus of the Branch, together with the warrant electricians (and those promoted therefrom), and a number of the civilian professional electrical engineers in the Director of Electrical Engineering's department. Some torpedo officers with valuable wartime experience in the operation of ships' electrical systems were also to be afforded the opportunity to transfer to the new Branch. Similarly, a number of electrical and radio mechanics were to be given the opportunity to transfer to regular engagements.

The Middleton report was not completed until after the end of hostilities and so steps were taken as a matter of expediency to call for volunteers from the RNVR electrical officers and others to turn over to the new Electrical Branch in late 1945. The new branch was set up under the direction of Rear Admiral S.L. Bateson on 1 January 1946, on which day the first list of RN electrical officers was published in Fleet Orders.

It is interesting to look in hindsight at the Middleton Report and observe how accurately the needs of the new branch were assessed and catered for by Admiral Middleton's committee. It was implemented almost without the need for modification, either in principle or in detail, surviving unchanged until the technology revolution brought about the Engineering Branch and Warfare Branch Developments. From Middleton we have grown the very satisfactory organisation for providing and supporting the latest technology in the highly complicated electrically based weapons, communications and command and control systems of today's Royal Navy.

CHAPTER 8

THE ELECTRICAL BRANCH

The Announcement

Against the backdrop of demobilisation and the trauma of six years at war, Admiralty Fleet Order (AFO) 7526/45 was issued promulgating the formation of a new Electrical Branch. A signal was released from the Admiralty at 1230z on 17 December 1945.

1. The new Electrical Branch, Royal Navy will come into being at 0001 on the 1st January 1946.

2. The existing Electrical Branch RN is transferred to the new branch on that date.

3. The List of Officers accepted for transfer to the (L) Branch will be promulgated forthwith, together with their seniorities and the arrangements for securing confirmation in the RN.

4. It is confirmed that the pay of the (L) Branch will be on engineering scales at the rate payable to electrical officers RNVR.

5. The first list of promotions to Commander in the new branch will follow in the first few days of the New Year.

6. At the Admiralty, the administration of the new branch, pending formation of an Electrical Department, will be coordinated under the Naval Assistant to the Second Sea Lord.

7. Existing responsibilities, except as modified above, will remain unaltered until further notice.

8. Further information will be promulgated as soon as possible.

The Branch Candidates

Throughout the Royal Navy the number of regular personnel readily available to be transferred into the new Electrical Branch on 1 January 1946 was relatively low. At the time, there were few RN electrically trained officers in the service who did not have the burden of either executive or marine engineering duties, thus making them suitable for the Electrical Branch but requiring voluntary or directed release from their parent branch. This left the warrant officer electricians and commissioned from warrant electricians as the most senior source of transferees. The bulk of the transferees were the existing regular service EAs who had been around since the early 1900s. They were to form the core expertise of the Electrical Branch and, under the new structure, they were to bring with them their existing equipment and maintenance responsibilities where there was a clear mandate from the Admiralty Board instruction. Apart from a departmental management change, the existing EAs would, both technically and in general, be largely unaffected by all the turbulence of restructuring. The existing personnel also broadly defined the initial establishment numbers of EA(L)s required for the new branch. However, this situation did change shortly after the war ended when the Electrical Branch finally took over the maintenance of all electrical generation equipment in the machinery spaces from the Engineer Branch.

Although not regular service ratings, the existing EA(R)s were capable of being transferred into the new branch. The electrical mechanics and the radio mechanics were full time technicians and, prior to demobilisation, they could be transferred as electricians and radio electricians. Some of these men were also considered to be suitable for artificer selection, particularly the radio specialists, but in view of the additional training required to bring them up to the standard of their peer artificers, they would be required voluntarily to transfer to a regular engagement in order to qualify for artificer status.

Although the career entry EA(R) was slowly working his way through the wartime apprentice training pipeline, he was unlikely to appear in the Fleet before 1948. This raised concern over the potential loss of radio and radar expertise to the Fleet which during the war had comprised of RNVR radar officers, direct entry EA(R)s and radio mechanics. One way of mitigating the problem was to offer the regular service EA(L)s the opportunity to cross train in radio engineering and this option was taken up by a significant number.

Although arguable, because they belonged to the Torpedo Branch, other ideal candidates for transferees were the LTOs in the low power sections on board ship. These men had dedicated electrical responsibilities and appeared to have little in common with the proposed underwater warfare-focused TAS Branch, details of which still had to be determined. However, the fact that in early 1946 the decision regarding the introduction of a TM Sub Branch and the torpedo artificer had not yet been reached, made directing the transfer of the LTO(LP) into the Electrical Branch more contentious in case they were needed as TM specialists in due course.

The Volunteers

RNVR officers were one of the prime target groups for recruitment into the Electrical Branch. Not only did they have the professional qualifications required of the new electrical officer, they also had wartime experience of radio and radar engineering. The issue was persuading such officers to give up a career in industry for a full-time career in the Royal Navy. Other potential volunteers were the specialist qualified executive officers who had some management experience of electrical technology, often with the assistance of the RNVR officer, and were prepared to give

"YOU CHAPS DONT SEEM TO REALIZE, THAT IF YOU ARE SMART AND EFFICIENT YOU MIGHT BE TAKEN ON AFTER THE WAR."

DINK

(Collingwood Museum)

up their executive aspirations and to undertake the additional technical training needed for them to qualify as electrical officers.

One surprising source of volunteers came from the Army. During the early years of the Second World War the British Army had realised that their existing repair system was not able to support the massive scale of equipment being deployed in every theatre. In 1941, the War Cabinet directed Sir William Beveridge to carry out a study on this issue and this resulted in the formation of the Corps of Royal Electrical and Mechanical Engineers (REME) on 1 October 1942. Desperate to recruit graduates into the new corps, the Army set up an engineering cadetship scheme with 4000 places filled in 1943, the first year of the scheme. (See Lieutenant Commander Peter Bates' account in Appendix 1.)

However, the end of the war quickly approached and the Army only took on some 400 engineering recruits. As a result, on graduating many of the initial entry of REME cadets volunteered to join the Royal Navy as electrical mechanics. During 1946, the Army officer cadets in the training pipeline, who by now knew that they had less chance of being taken on by the REME, were invited to apply for commissions in the Royal Navy as electrical officers. The fact that the REME training course had been recognised by the Institution of Electrical Engineering was a bonus as it was exactly the professional qualification envisaged for the new branch.

Another prime source of volunteers was DEE where many of the civilian engineers belonged to the Admiralty Electrical Engineering Service and had relevant naval experience. They also had the professional qualifications expected of the new electrical officer. The fact that DEE could possibly become 'navalised' at some stage after the birth of the Electrical Branch would also have been a motivating influence on the decision to volunteer for the Royal Navy.

The Middleton Report had expressed some concern about the supply of volunteers given the turbulence of demobilisation and potential sources of electrical expertise being embedded in the existing branches. The latter was clearly a point to be seriously considered as the other branches were also seeking to develop their own structures given the increasing complexity of weapon systems. The Engineer Branch had recognised for some time the similarity in engineering skills needed for ordnance engineering functions and had invited some of its officers to sub-specialise in gunnery-related engineering. Eventually, this led to the Ordnance Branch being subsumed into the Engineer Branch and the ordnance officer's dark blue specialist stripe was replaced by the purple of the engineer.

AFO 517/45, issued on 1 February 1945 to give a preliminary announcement of the formation of the Electrical Branch and encourage volunteers, had clearly made a favourable impression on many and volunteers did start to come forward. By March 1947, 52 regular service torpedo officers and 13 communications officers had volunteered to join the Electrical Branch along with 40 warrant officer telegraphists and 120 warrant gunners (T).

The fledgling nature of the Electrical Branch was clearly a concern to some senior officers and Commander Cyril Locke tells of a letter written by Admiral Mountbatten to the Fleet Electrical Officer which encapsulated some Executive Branch concerns (see Appendix 3 for details).

(Collingwood Magazine Archives)

HMS *Collingwood* – The Lead Electrical Training School

During the war, electrical training had been carried out in many establishments throughout the country and the introduction of the Electrical Branch was a clear opportunity to rationalise the number of sites and at the same time update training capability to support the many systems which had been fitted in the Fleet during hostilities. Following the recommendations of the Phillips Report, and subsequently the Middleton Steering Committee, *Collingwood* was selected to be the Royal Navy's lead electrical training school with special responsibility for the Portsmouth home port area. The other port area schools nominated were *Defiance* for Devonport and *Actaeon* Building[107] at Chatham. The training in these schools was be coordinated by *Collingwood* under DNLD. HMS *Ariel*, the naval air station at Warrington, was selected as the electrical training school for the Fleet Air Arm but still under the training coordination of *Collingwood*.

The case for selecting *Collingwood* as the lead centre for Electrical Branch training was prompted by its existing training infrastructure and its capacity to absorb a large number of trainees. While the utilisation of existing facilities offered significant cost savings, another benefit was the fact that the establishment already had many of existing electrical training facilities, including radio and radar instructional equipment which had the freedom to carry out the live transmissions needed for operator training and essential for maintainer training. Notwithstanding its existing capacity, with the imminent arrival of the electrical artificer and the electricians, a major building programme was started to provide extra classrooms, a bigger radio section and general electrical training facility which later became a famous landmark known as 'the White City'. This was because buildings constructed during the war were camouflage painted but as this was no longer necessary, the new building's walls and roofs were left with a white finish.

At Chatham, *Actaeon* Building was only given limited electrical training facilities in 1946 but *Defiance* eventually acquired a capability for general electrical, radar, radio and sonar technical training as well as basic training of ratings called up into the Royal Navy under National Service.

Most other training facilities from around the country, including those electrical training facilities evacuated from *Vernon*, were eventually relocated to *Collingwood* and by 1947 it was established as the Electrical School of the Royal Navy. During this period *Collingwood* also acted as a post-

HMS *Collingwood* Electrical School parade ground, 1946. (Collingwood Photographic Department)

HMS
COLLINGWOOD
COMMITTED TO
EXCELLENCE

(Collingwood Magazine Archives)

war demobilisation centre and continued to carry out some seamanship training using the facilities installed at the beginning of the war. However, by 1948 seamanship training had been phased out to other shore establishments such as HMS *Excellent* and HMS *Dryad*. In August 1947, Captain Gerald Jackson became the first electrical officer to take over command of *Collingwood*. Captain Jackson had been in command of HMS *Marlborough*, the wartime electrical training centre, from 1942–1947.

Establishing the Branch

The work to set up a cohesive branch soon began in earnest and by March 1947, 253 RNVR specialist officers and 50 warrant officers had been given commissions in the Electrical Branch. All of the officers required some degree of cross training in order to meet the full criteria for an electrical officer and this was well underway in *Collingwood* by 1947.

Many ratings had started to enter the new electrical school with one branch badge on their arm and, after a completing an appropriate conversion course, left with another showing (L). The chief petty officer and petty officer torpedo gunner's mate became the chief electrician and petty officer electrician, retaining their previous titles in brackets until they had been successfully cross-trained. Similarly, the leading torpedo operator became a petty officer electrician. The 'Hostilities Only' rating of electrical mechanic was eventually discontinued under the demobilisation programme with some entering *Collingwood* to carry out the supplementary general training needed to hold the rate in a regular service engagement as a chief or petty officer electrician. The seaman torpedomen who volunteered to transfer were re-categorised into one of the new rates in the Electrician's Mate structure depending on skills and experience. New entry recruits were brought in as Probationary Electrician's Mates (PEMs) before specialising in general electrical or radio engineering. Subsequently their advancement was through EM second class (EM2), EM first class (EM1) to Leading Electrician's Mate (LEM) before being eligible for advancement to petty officer electrician. The EM junior ratings were classified as semi-skilled ratings and their training up to and including leading rate was very much focused on ship equipment maintenance tasks of a general electrical, radio or radar nature.

In 1955 Their Lordships issued AFO 824/55 which announced that the titles of Electrical Branch ratings were 'giving persons unfamiliar with Naval tradition and recent developments onboard ship the wrong impression of the duties performed by the Branch.' To clarify the matter it was decided that the junior rate title of Electrician's Mate should be changed to Electrical Mechanic with effect from 14 March 1955. For the senior rates, the title of Chief Electrician was to remain and the Electrician was accorded the title Petty Officer Electrician. Similar changes were implemented for the radio specialists, giving rise to the Petty Officer Radio Electrician and the Radio Electrical Mechanic. For the same reasons, and at the same time, the Stoker was re-categorised as an Engineering Mechanic.

The Series Apprentice Training Scheme

The wartime recruitment of skilled mechanics from industry to fulfil electrical and radio maintenance roles had created a source of specialists who started to leave the Service under demobilisation. It quickly became apparent that there was going to be a skills gap, particularly in the radio and radar fields because of the immediate shortage of any regular Service-trained EA(R)s. Some of the shortage was taken up by 'Hostilities Only' radio mechanics and chief petty officer and petty officer telegraphists who had volunteered to join the Electrical Branch as radio electricians. Despite these volunteers, the situation still necessitated a crash training programme for selected EA(L)s who had transferred to the new branch and volunteered to plug the gap until the first of the new EA(R)s were ready to join the Fleet.

During the autumn of 1946, the Engineer-in-Chief ordered a review of the whole of naval apprentice training. The object of the review was to submit proposals for future artificer apprentice training with a view to bringing as much electrical training under the control of electrical officers and to reduce the length of the training time, which for an EA(R) was approaching six years. A meeting was held at the Admiralty on 15 November 1946 to discuss the proposals. At the meeting, the decision was taken to introduce a new system for all apprentice training which was to be called the 'Series Scheme'.

While the previous system had consisted of two entries of artificer every year, the Series Scheme was to comprise three entries or classes per year which aligned with the three school terms in the academic year. The new scheme was put into effect with the apprentice entry of August 1947 and the apprenticeship was restructured to be completed in four years, divided into twelve terms and numbered accordingly. The reduction in training time was achieved by reducing the amount of craft training to be commensurate with the new technology. Under the Series Scheme the selection for all apprentices was carried out in two stages, the first being an educational screening at various centres around the country. This first stage took about a day and the successful applicants then proceeded to a second stage which was a two-day visit to either *Ariel* for northern based recruits or *Collingwood* for those living in the south. During the two days the candidates were subjected to aptitude tests and interviews. The standards set for selection were very high and rigorously applied and this led to around 40 per cent of candidates interviewed being declined.

Those who successfully passed the second stage were allocated a place at HMS *Fisgard*, the new entry training establishment founded at Torpoint in Cornwall. The first three terms were spent at *Fisgard* where the apprentices learnt general trade skills and undertook instruction that was common for all artificer specialisations. After the first common year, the apprentices were categorised based on aptitude and those selected for the electrical artificer apprentice stream did a fourth term at *Fisgard*, which had a more electrical bias to the training syllabus, before moving on to *Collingwood* for their fifth term.

The first series class of electrical apprentices joined *Collingwood* in September 1948. Term 5 was spent undergoing instruction in basic electricity and electronic theory and workshop training in the skill of hand needed to carry out practical fitting and turning. At the end of this term, again based on individual aptitude, the class was divided into two streams and allocated for further training as either an EA apprentice or a Radio Electrical Artificer (REA) apprentice.[108] The allocation to specialisation was approximately 60 per cent EA and 40 per cent REA. During terms 6 to 9, the apprentices studied electricity and electronics in depth and their training programme included more practical workshop experience. In both cases the training was geared to the technology to be found in their specialist areas. After term 9 there was a further subdivision to select apprentices for Fleet Air Arm training as an EA(Air) or an REA(Air). After selection, the air electrical specialists went on to join *Ariel* and complete their training, while the EAs and REAs remained at *Collingwood*. During the last year of their apprenticeship at *Collingwood*, the EA and REA apprentices were given detailed training in specific ship equipments, electrical or radio and radar systems, as appropriate, prior to being sent to their first operational sea draft.

It is of interest that in 1948 the term 'conscription', which had been reintroduced in 1939 by the passing of the National Service (Armed Forces) Act, ceased to be used and was replaced by

Series I apprentices passing out, class of 1951. (Collingwood Museum)

HMS *Defiance* Electrical School, Devonport. (Collingwood Magazine Archives)

the term 'National Service'. Under the conditions of National Service there was an EA(National Service) entry and these men were trained as artificers in *Defiance*, not *Collingwood*. In a *Live Wire* newsletter from *Defiance*, published in 1949, it was reported that the numbers under instruction had continued to rise, particularly the EA(National Service) trainees. As the Royal Navy reduced its manpower to suit the smaller post-war navy, so the number of conscripted artificers reduced and this eventually contributed to *Defiance* closing as the Devonport Electrical Training School on 14 July 1954.

The Arrival of the Mechanician

From very early in the post-war recovery period the electronics industry was growing rapidly and skilled manpower became very scarce. Recruiting was made difficult particularly for the Electrical Branch, which was itself expanding to keep pace with new developments in the technology. The recruitment of direct entry artificers was reduced to a trickle by the tempting opportunities available in industry and the number of artificer apprentices being produced under the old apprentice scheme was lagging behind the manning requirement. It was hoped that the new Series Scheme which started in 1947, would encourage more recruits, but the scheme was not due to provide the 'new' artificer into the Fleet until 1951.

In 1944, the predicted shortage of artificers due to enter the Fleet and the fact that the 'Hostilities Only' REM and EM ratings would soon start to leave under demobilisation, highlighted a need to bridge the skills gap between the artificer and the new electrician rates. The Phillips Committee had foreseen this and their proposals had been to introduce a new electrician rate who would, in a similar manner to the wartime EM, be directly recruited into the Electrical Branch as an intermediate skill level maintenance resource between the artificer and the electrician's mate.

However, after reviewing the Phillips Report, the Admiralty rejected this proposal and instead advocated following the example of the Engineer Branch by the introduction of the non-substantive rate of mechanician. The mechanician rate was to apply to both electrical and radio specialist streams of the Electrical Branch but the personnel were to be selected from the most able of the new regular service EM and REM ratings who had reached the acting leading rate or above. Selection for a mechanician's course was to be based on the marks achieved on the qualifying course for leading rate, followed by suitable recommendations from sea. Qualified leading ratings and above who were serving at sea could also be selected for the course based on their performance in a sea-going billet. After final selection, the electrician or electrician's mate returned to *Collingwood* to undertake a two-year mechanician's course in order to qualify them up to the Artificer Third Class non-substantive rate which then enabled them to be drafted to ships as either an electrical or radio mechanician in lieu of an artificer.

A major benefit of the mechanician concept was the fact that it provided the electrician's mate with access to a higher technical qualification and this acted as an incentive to stay in the Service. Perhaps more importantly, it now opened up more promotion opportunities for those electrician's mates who had not been able to meet the entrance qualifications for artificer and not been selected on entry for accelerated promotion to commissioned officer.

The first mechanician's course was completed in 1954 and the men joined the Fleet in the summer of that year. Most of the early qualifiers were experienced petty officers selected from sea, who had welcomed this opportunity to advance themselves. The other source of candidates came from direct entry recruits who had a suitable industrial electrical background and experience. Many of the direct entries were men who had been too old to take the artificer apprentice entrance exam and gone on to take civilian apprenticeships but were now looking to the Royal Navy for a career. The motivation of both of these groups was unquestionable and they brought a wealth of practical skills and common sense into the classroom.

After qualifying as a mechanician, the advancement opportunities opened up to allow promotion not only to chief radio or electrical mechanician, equivalent to the artificer counterparts, but also, if successful in the professional examination and selection process, the man could now become a commissioned electrical officer (L) or (R). In many cases the maturity of the mechanician was such that his age and experience made him eligible to take the exam for a commission comparatively quickly after his qualifying course.

The success of this scheme over the next few years, coupled with the decline in the numbers of artificer apprentices coming into the Service, was no doubt instrumental in the alignment of the Mechanician Scheme with the Artificers Series Scheme in 1964, when many of the training course modules become common to both groups and the mechanician apprentice came into being. Eventually, in 1971, DCI(RN)1140/71 announced that the artificers' and mechanicians' training schemes would become fully aligned. The upper age limit for artificer entry was extended to 25 in order to match the maximum age of eligibility for mechanician and give the more mature recruit, without any industrial training, the opportunity to enter as an artificer.

Changes to the Officer Structure

The Warrant Electrician and Commissioned from Warrant Electrician ranks had been in existence since 1911 but with the formation of the new Electrical Branch they were changed to Warrant Electrical Officer and Commissioned Warrant Electrical Officer. These ranks were followed by (L) or (R) to reflect specialist qualifications and provide an equivalent promotion structure for the new breed of EAs and REAs coming through the training system. However, following a report by an Admiralty committee on the status of the warrant officer led by Admiral Noble, the warrant rank was abolished in 1949. The old ranks were replaced by the new ranks of Commissioned Electrical Officer and Senior Commissioned Electrical Officer, followed by the specialist suffix (L) or (R). The Noble Committee also set up the Branch List which comprised all commissioned officers who were ex-warrant officers, abolished separate warrant officers messes and gave all commissioned from warrant officers wardroom status.

In early 1955, it was decided as a precursor to much wider changes to the officer structure to dispense with the coloured branch cloth that specialist officers of the Supply, Instructor, Electrical and Engineer Branches had been required to wear between their gold rings since 1925. Thereafter, the only active duty branches to continue wearing the coloured cloth to denote their profession were medical officers, including ward masters and dental officers. Thus the wearing of the 'green' by the new Electrical Branch had lasted less than ten years although, as the title of this book shows, the tradition lived on.[109]

A New Officer Structure – AFO1/56

The more significant changes to the officer structure appeared with the promulgation of AFO 1/56, which announced the creation on 1 January 1957 of a new General List (GL) that embraced all Executive, Supply, Electrical and Engineering Lists. At the same time the Special Duties (SD) List and the Supplementary List (SL) were created in recognition of the fact that the GL entry numbers

(Collingwood Magazine
Archives)

AND WHICH IS THE ELECTRICAL OFFICER ?

were to be reduced to enhance promotion prospects to commander and above, thus leaving more commissioned officer billets available at the junior level to be filled from the other lists.

The GL was to be based on direct entry at the age of 18, specialisation after a period of common training and a career expectation of up to 55 years of age depending on rank. The issue of eligibility for command was addressed in the announcement and it was confirmed that only executive officers could achieve sea command. The concession offered in return for this exclusion was that other GL officers were given general powers of military command, which excluded sea command, and this allowed them to take command of a shore establishment or non-seagoing ships. For the electrical officer, this latter concession was not totally new as the precedent for shore command had already been set by Captain Jackson, who took command of *Collingwood* in 1947.

The Branch List, a term which had never impressed many of those who appeared in it, was renamed the Special Duties List. The SD List became a list of officers promoted from the lower deck and these officers were differentiated from the GL only to indicate the specialist nature of their technical training and experience. All Branch List officers were transferred to the SD List and changed their rank from Commissioned Electrical Officers to Electrical Sub Lieutenants, exchanging their single half-width stripe for a full width stripe. Senior Commissioned Electrical Officers became Electrical Lieutenants. Under the Order in Council CW2032/1920 dated 22 January 1920, a few electricians had already been promoted to electrical lieutenant and these officers were immediately made up to become Electrical Lieutenant Commanders with an additional half stripe. All officers on the SD List retained their sub-specialist suffix of (RE), (L), (AR) or (AL).

The Supplementary List was designed to accommodate direct entry officers on short service commissions, up to 16 years, who were required to fill temporary complement shortages in GL manning levels. The initial recruiting requirement for this list was for aircrew and electrical specialists only.

Shortly after this restructuring, Their Lordships announced in AFO3023/58 that they were to set up a committee, chaired by Vice Admiral Sir Stephen Carlill, to examine the future responsibilities of the Electrical and Engineering Branches and professional training of GL engineer officers. The report was published in January 1960 and recommended, in principle, a single engineering specialisation for which all GL technical officers would receive common electrical and mechanical professional training before sub-specialising into either the Electrical Branch or the Engineering Branch. A Steering Committee was appointed in 1960, led by the CNEO, Vice Admiral Sir Norman Dalton, to oversee the implementation of the changes proposed by Carlill. The Engineering Specialisation Working Party (ESWP) was formed to work under the direction of the Steering Committee.

(Collingwood
Magazine Archives)

The Control Specialist

The onset of the integrated combat system in the 1960s and the wider use of electrically driven gun mountings and missile launchers, produced advances in control engineering techniques being used onboard ship. Rapid response weapons were operationally required and, with the emergence of computers, multiple fire control channels were becoming available to drive those weapons. In 1961, Their Lordships decided that the existing artificer structure was insufficient to allow the deep specialist skills needed for this expanding area of technology and a new rate of Control Artificer(Weapons)(CA(W)) was introduced. The CA(W) was to be given responsibility for both surface and underwater fire control systems and the remote power control systems associated with gun mountings and missile launchers. As such, the CA(W) was to provide the skilled technical expertise needed to directly support the users in the operation of the weapon elements of the ship's combat system. Effectively, this scope of employment involved taking over the control engineering duties then being undertaken by the EA, Electrical Mechanician and the OA.

In conjunction with the rise in control engineering, there was also a view that the requirement for ordnance engineering per se was diminishing due to a perceived reduction in the nature of the ordnance support required for the future Fleet due to the loss of the Capital ship. In addition, the acknowledged success of the Electrical Mechanician in filling high power generation and distribution roles in ships gave rise to the idea that these billets could be designated as Mechanician billets. It was then postulated that, over time, the effects of the ordnance technology trend, mechanician complementing action and the rise of the CA(W) would eventually remove the need for the EA and OA specialists. Naturally, there were mixed reactions as to the potential consequences of introducing the CA(W) but before these could be tested, on 24th April 1961 in AFO650/61, the planned amalgamation of the Electrical and Engineering specialisations was announced. This was to be achieved in the future by the Engineering Department assuming responsibility for electrical power generation and distribution and, in the interim, the appointment of Electrical Branch SD(L) officers into suitable seagoing electrical posts where the Engineer Officer was deemed professionally competent to take on electrical responsibility.

CHAPTER 9

ENTERING THE COMPUTER AGE 1946–1966

The Future Threat

The V1 and V2 rockets are notorious in terms of the loss of civilian lives, but the full extent of the advanced nature of the German missile development programmes is less well known. As such, while it came too late in the war for the German Reich, it is worth recapping what nearly might have been and how the Allies' glimpse of the emerging German missile technology would have been factored into the Royal Navy's post-war ship and weapon equipment strategy.

Apart from the appearance of the first German jet fighters towards the end of the war, another prime consideration would have been the future anti-ship missile threat. The latter had started to appear in August 1943 when the German HS 293 air-launched, rocket-propelled 'Glider Bomb', or anti-ship missile, was first used operationally against the Allies off Finisterre. In an effort to find out more about the weapon, HMS *Egret* was sent out as part of a decoy group to try and lure the Germans into attacking with the HS 293, so that its nature could be observed. The ploy succeeded but unfortunately *Egret* was hit by one of the bombs and sank with the loss of 194 lives including naval personnel and a considerable number of scientists embarked for the mission. Observations during this action discovered that the operational HS 293 was powered and radio controlled under visual guidance from an aircraft. After the war it soon emerged that a camera with a video retransmission facility was on the drawing board to be fitted to the HS 293 and this would have given a blind-fire capability not even imagined pre-war.

A similar air-to-ship threat was posed by the German FX1400 missile. This missile was more of a guided armour-piercing bomb that used gravity to generate sufficient air flow over remotely controlled tail fins which were the operated to guide it to its target. The FX1400 was used at the Battle of Salerno in September 1943 when HM Ships *Spartan* and I*nglefield*, as well as two Liberty ships, were lost due to what became known as the Fritz X bomb. HMS *Warspite* was also hit by three FX1400 bombs during the action but survived despite one bomb piercing the hull.

The emergence of both guided weapons led to immediate countermeasures attempts by the British and it was probably these events that started the expansion of the Royal Navy's RCM interests to include not just confusion and information denial but also ship point defence; eventually this field of applied technology became referred to as 'electronic warfare'.

The other significant, post-war revelation about the German missile programme was the existence of the 'Wasserfall', a supersonic anti-aircraft missile system which had a design performance requirement of being able to intercept a target flying at 560mph up to a height of 65,000 feet. This system used one radar to track the target and another to track the missile with a command facility to bring the two tracks into coincidence, in other words the first conception of a command line of sight missile system. It was also thought that both a homing system and a proximity fuze were being developed for this missile.

While the scope of the potential missile threats raised new questions about maritime warfare tactics and ship defensive systems, the known U-boat threat was also reassessed in the light of the

German technology captured at the end of the war. As early as July 1945, Flag Officer Submarines, Rear Admiral G.E. Creasy, at last felt able to admit the opinion that 'undoubtedly, we led the Navies of the world in the detection of the submerged submarine, though I believe we tended to exaggerate our abilities in this respect and we were certainly over-optimistic in our estimates of our powers to "kill" once we had "detected."'

With these wartime lessons clearly in mind, the future prospect of having to deal with a fast, deep-diving submarine with the endurance to complete a patrol fully submerged had to be seriously considered. In common with other Allies, the Royal Navy closely examined the latest German hull designs and the capability of their passive sonar systems was analysed. Along with the US and the Russians, the Royal Navy finally started to take interest in the submarine as an ASW platform and the concept of 'Hunter Killer' submarine operations came into being.

The Future Technology

Looking back there have been significant advances during the history of electrical technology which have opened up further potential uses in the naval warfare environment, bringing new areas of military capability as well as new areas of threat. Arguably, the advances of most significance to the Royal Navy were the invention of the battery in 1799 leading to its use for firing circuits; the dynamo in the mid-1800s being used for powering searchlights; the coherer radio wave detector in 1890 with its use in radio communications; and the thermionic valve in 1906, which towards the end of the First World War led to the explosion in radio and radar technology both up to and during the Second World War.

While these advances generated new operational capability for the Royal Navy, they also represented landmark advances in technology which required step changes in the way in which that technology was managed. In brief, it seemed that each new advance was destined to bring about a significant change in the way the Fleet's manpower needed to be structured.

The next significant change actually started to emerge before the war at the Bell Laboratory in the US where what was known as the 'semiconductor' properties of certain materials were finally explained. The basis for the explanation was that silicon was found to have positively and negatively charged ionic regions and that this property could have possible electronic applications. This farsighted conclusion proved to be the basis for the next fundamental step change in electrical technology, which was to have an even greater impact on advances in naval capability than any of the previously mentioned landmarks.

It was in December 1939 that a scientist, William Shockley, viewed the research findings at the Bell Laboratory and postulated the use of semiconducting materials in electronic amplification circuits. This offered the possibility of replacing the functionality of the fragile radio valve with a solid state device which was far more robust.

However, it was only after the war that Shockley was able to put together a research team comprising John Bardeen and Walter Brattain to work on his ideas. It was not until 23 December 1947 that Shockley's team was finally able to demonstrate the first semiconductor device. This experimental device was actually made using germanium, a material which had similar semiconductor properties to silicon.

The name given to the device was the 'transistor', a contracted combination of the word transconductance and varistor[110] and, in its demonstrated form, it had the same functionality as a triode valve. Out of this initial configuration grew a family of transistors that found applications in both the military and civil environments with one of the first being the eponymous transistor radio, followed closely by the digital computer. Thus the stage was set for a technological shift that would require close monitoring and timely responses by the Royal Navy in order to adapt both its technical training and management approach.

Post-War Events

The final defeat of Japan on 14 August 1945 saw the Royal Navy in its largest form ever. It comprised 14 battleships, 52 aircraft carriers, 257 destroyers, 131 submarines and 9,000 smaller vessels. The manpower level had also been increased to 865,000 from the 134,000 level in 1939. However, the cost of the war had been enormous in both material and men and the country was heavily in debt to the US and nearly bankrupt. By 1947, the manpower level had been reduced to 195,000 as 'Hostilities Only' personnel and reservists left the Service. The reduction in the size of the active Fleet was commensurately reduced with hundreds of ships being placed in reserve awaiting eventual disposal.

Notwithstanding the reductions in the Fleet, technology continued to advance and the next 20 years saw the introduction of improved radar and sonar systems and the arrival of the computer and early attempts to integrate weapons and sensors to form a combat system. Electrical technology started to embrace scientific advances and the move began from valve systems to solid state systems. The operating characteristics and reduction in size of the new components now gave opportunities to put more effective capability into smaller hulls. These advances can best be illustrated by recalling the evolution of the frigate escort.

The Frigate Escort Programme

Following studies in 1947, it was decided that it would be more strategically and cost effective to build escort frigates using a design which facilitated prefabrication and allowed the fitting of weapon and sensors tailored to meet specialist warfare requirements. This thinking was driven by the onset of the nuclear threat giving rise to the perception that shipbuilding facilities should be spread throughout the country in order to minimise the potential damage that might result from a nuclear strike. Essentially the policy adopted was that the hulls would be built with the same propulsion and machinery systems and that weapons and sensors would be installed to match the operational capability required. The common platform concept was put into place when the Admiralty approved the 1951 frigate programme. This programme was to include the Type 41 Leopard Class anti-aircraft frigate, the Type 61 Salisbury Class air direction frigate and the Type 11 anti-submarine warfare frigate.

The Type 41 Leopard Class

Four Type 41s were laid down in 1953 and on build they were fitted with forward and aft mounted 4.5in Mk6 turrets and two 40mm Bofors mountings but not Sea Cat, a close range guided missile system and the only one available with an anti-aircraft capability at the time. Unfortunately, the 4.5in gun had only a very limited capability at medium range in anti-aircraft mode, while the Second World War Bofors guns were of limited effectiveness against the emerging threat of supersonic aircraft and missiles.

The ships were fitted with asdic Type 164 attack set and Type 174 passive warning set. Both sets were still reliant on wartime technology with the Type 164 linked to a single forward firing anti-submarine squid mounting. The top speed of some 24 knots meant that the Type 41s were of limited use in ASW operations particularly with the imminent arrival of the much faster nuclear submarine. However, there were two redeeming features for the class. The first was that they had a diesel engine propulsion system which offered long endurance and savings in manpower, but it was this factor which contributed to the ship's speed limitations. Secondly, the Mk6 was very effective in surface engagement and naval gunfire support modes, roles which were secondary when the design was conceived but eventually served a useful operational purpose. The Type 41 Class was never upgraded and the four ships of the class, HM Ships *Leopard*, *Lynx*, *Puma* and *Jaguar* had been taken out of service by 1975.

HMS *Lynx* Type 41 Leopard Class anti-aircraft frigate. (Naval Museum Portsmouth)

The Type 61 Salisbury Class

The first of the Type 61s was laid down in 1952 as an air direction frigate. Its role was to act as a radar picket deployed ahead of the Fleet to give early warning of enemy aircraft and to direct either shore based or carrier borne aircraft in the event of an attack. For this task the ships were fitted with the Type 982 air search radar in lieu of the aft 4.5in Mk6 turret fitted in the Type 41. As with the Type 41, speed was not considered to be of the essence for the operational task and the Salisbury Class also had a diesel propulsion system installed. Only four Type 61s were built before the concept of having multi-purpose frigates was implemented and all four ships had been retired from service by 1978.

HMS *Llandaff* Type 61 Salisbury Class air direction frigate. (Naval Museum Portsmouth)

The Type 12 Whitby and Rothesay Classes

The Type 11 was originally conceived as an anti-submarine frigate but it was eventually abandoned in favour of the Type 12 Whitby Class. The Whitby Class was still regarded as a specialist anti-submarine frigate and six ships entered service between 1956 and 1958, but as early as 1954 the cost and effectiveness of building specialised hulls was being questioned. Consequently, the Whitby design was updated to include a Sea Cat missile system and this design variant was designated as the Rothesay Class. The main armament for both classes also included a 4.5in Mk6 gun and a twin mounting ASW Mortar Mark 10 (Limbo) which gave an all round mortar launch capability out to 1000 yards. The radar suite comprised Radar 993 for medium range surveillance and target indication and Radar 978 for navigation. The target tracking radar, Type 903, was integral to the MRS3 (Medium Range System Mod 3) fire control director for the gun. Sonar sensors included Type 177 medium range search sonar, Type 162 bottom search and the Type 170 fire control sonar. Type 170 could establish submarine range, bearing and depth and this data was fed to the MCS10 fire control computer used to control the Limbo ASW mortar which had replaced the wartime Squid. The MCS10 computer was arguably one of the most complex analogue computers in service and it used variable potentiometers to model the flight pattern of six unsophisticated mortars through the air and water to their optimum submarine killing explosion point. The firing pattern comprised three mortars triangularly placed above and below the submarine's predicted position with detonation being depth controlled. Within the constraints of the weapon and sensor systems available, the Rothesay Class became a model for the multi-purpose frigate.

Type 12 frigates of the 6th Frigate Squadron. (Naval Museum Portsmouth)

Twin Limbo antisubmarine mortar mounting fitted to a Type 12 frigate. (Naval Museum Portsmouth)

Ammunitioning Limbo mortar Mark 10 projectiles. (Collingwood Museum)

The Type 81 Tribal Class

In addition to the Rothesay Class, a new design for a general-purpose frigate, the Type 81 Tribal Class was proposed. The first Tribal was laid down in 1958 and it was designed to operate in a multi-threat environment. The main armament included two MRS3-controlled single 4.5in Mk5 guns, recovered from the scrapped Cavalier Class destroyers, and a GWS 21 Sea Cat System, controlling two quadruple missile launchers with either aimer control or radar tracking control using the Type 262 radar. The ship had a small hangar built into the flight deck and the class was instrumental in developing small-ship helicopter operating procedures. The flight deck and hangar arrangement along with the aft gun took up additional space in what was a fixed size hull and, in consequence, only a single Limbo ASW mortar was fitted. The radar and sonar sensor outfit was much the same as the Rothesays and both classes had the facilities to operate a Wasp helicopter carrying an anti-submarine torpedo or the AS12 air to surface missile.

Another significant innovation brought in with the Type 81 was the propulsion system, which involved the first use of a Metropolitan Vickers G6 gas turbine in conjunction with a traditional steam turbine. These turbines could be used through a common gear box to drive a single screw. This configuration became known as a Combined Steam and Gas (CoSAG) propulsion system. The G6 was a marine version of an aircraft engine that could provide extra power on immediate demand, as it did not need the warm-through time of a conventional steam propulsion system. Although versatile, the gas turbine was inefficient at low speeds and this made the steam turbine the preferred system for cruising at speeds up to around 20 knots. With the gas turbine linked into the drive system, the top speed went up to some 28 knots. The introduction of the G6 gas turbine into an operational class was first trialled in HMS *Exmouth* and showed that the advances in technology at this time were not be confined to the area of weapons and sensors. In hindsight, the implementation in the Type 81 was inelegant, for while space was saved below decks in having less boiler room space, above decks the G6 required its own funnel. This not only led to a unique ship profile but it also constrained the deck space that could be made available for helicopter operations. However, more significantly for the Marine Engineering Department, the G6 turbine gave a glimpse of the changing nature of marine propulsion and its associated control systems. Whereas some thought that there was nothing to challenge steam, they were to be proved wrong and 'gas' technology was to become a major driver for organisational change later on in the twentieth century.

HMS *Ashanti* Type 81 Tribal Class general purpose frigate. (Naval Museum Portsmouth)

Seacat missile firing. (Naval Museum Portsmouth)

Wasp helicopter small-ship operating trials on HMS *Nubian*, 1963. (Fleet Air Arm Museum)

Wasp Type 81 hangar
stowage trials on HMS
Nubian, 1962. (Fleet Air
Arm Museum)

From a personnel standpoint, the Type 81 was a pioneering design. It was one of the first classes to be fully air conditioned throughout the ship and it heralded the introduction of cafeteria-style messing and bunks, rather than hammocks, for everyone on board.

The Type 81 was designed as a multi-purpose frigate but did not have the equipment fitted to allow it to undertake any operational task well, with performance limitations being accepted at the expense of versatility. The single Limbo could not deliver the optimum killing pattern for the AS mortar system and the 4.5in Mk5 guns were even less effective in anti-aircraft mode than the 4.5in Mk6 twin mounting fitted in the Type 12s. One advantage, noted by the ships' companies, was the fact that the domestic air conditioning introduced made it very suitable for service in the Middle East. However, as was discovered during the Cod Wars of the 1970s, the single screw was also a severe manoeuvring limitation when nimble Icelandic gun boats made life very difficult for the Type 81 in its fishery protection role.

Conventional Submarines

In 1948, eight of the latest T Class submarines, which were the first to have all-welded hulls, were put though an extensive 'Super-T' conversion at Chatham dockyard. The modifications included the removal of deck guns and the replacement of the conning tower with a 'sail', a smooth-surfaced and far more symmetrical and streamlined conning tower. A new section of hull was inserted to accommodate an extra pair of electrical motors and switchgear and an extra battery was installed. These changes increased the underwater speed up to 15 knots or more and the underwater endurance to around 32 hours at 3 knots. The 'Super T' Class conversions were also fitted with more complex electrics and electronics, including advanced sonar and much

HMS *Exmouth* G6 gas turbine trials ship. (Naval Museum Portsmouth)

HMS *Falmouth* Rothesay Class returning from a Cod War collision with the Icelandic gunboat *Týr* in 1977. (Private Collection)

HMS *Porpoise* in 1961. (Naval Museum Portsmouth)

improved torpedo control systems. The first boat to be modified was *Taciturn* in March 1951 and the programme was completed with the conversion of *Trump* in June 1956.

In December 1950, approval was given for the streamlining of five of the earlier T Class that had been built with riveted hulls. These boats were not to have the hull extension but the main batteries were replaced with more modern and efficient versions which provided a 23 per cent increase in power. This decision meant the conversion was less complex and costly and the work could be undertaken during a normal refit. The first riveted boat to undergo this conversion was *Tireless* in 1951.

Following the increase in on-board sensor technology and the on-board experience of RNVR electrical officers during the war, it was decided to appoint Electrical Branch officers additional to the submarine's normal complement. The object was to gain experience in the employment of such officers at sea in submarines with a view to providing the new Porpoise Class with a complement electrical officer, instead of the customary fifth executive officer. For this appointment the electrical officer had to undertake full submarine training and qualify for submarine pay before taking up the billet. Apart from his electrical duties, he could be required to undertake watchkeeping duty on the surface, snorting or dived, and to participate as a member of the attack team. Despite his executive training and duties, the submarine electrical officer, like his General Service counterparts, could still not aspire to sea command.

The Porpoise Class was a radical departure from previous hull design practice, taking some design improvements from German U-boat technology. The boat was also fitted with a high-voltage propulsion system with batteries connected to give 880 volts for maximum speed and 440 volts for patrol speeds. Mechanical operations at these voltages presented switchgear design challenges that were resolved by introducing motor-driven contactors with fault protection facilities operating at high speed and with a high breaking capacity. In order to develop this switchgear the Admiralty engineering laboratory had to build one of the largest DC short circuit testing installations in Europe at the time.

The Porpoise Class design was exceptionally quiet underwater, more so than other NATO submarines and much better than the Soviet Whiskey Class, which was providing the opposition. This was in part due to careful attention to detail in the mounting of machinery and advances in propeller design that prevented cavitation noise. The silent running capability made the submarine a particularly effective sonar operating platform and ideal for Special Forces operations. The class was also the first not to have a gun mounting on build and this policy was adopted for all follow on classes of Royal Navy submarines.

Nuclear Submarines

In addition to the technological improvements in conventionally propelled submarines, the early 1960s saw the first use of nuclear propulsion, which was to enable the deployment of ballistic missiles in submarines as the Royal Navy's contribution to the UK's national defence strategy. What gas turbine technology had done for the marine engineering departments in the Surface Fleet, nuclear propulsion was doing in the Submarine Fleet.

The Royal Navy's first nuclear attack submarine (SSN), HMS *Dreadnought* was built at Barrow and fitted with a US-built pressurised water reactor. The propulsion system and the hull design were based on the US Skipjack Class. *Dreadnought* was commissioned in 1963 and three years later the Royal Navy's first nuclear strategic submarine (SSBN), HMS *Resolution*, came into service. The submarine was fitted with the Polaris ballistic missile system and formed part of the nuclear strategic deterrent.

Apart from the nuclear propulsion, both SSN and SSBN hull forms incorporated new designs that offered significant improvements in the speed, endurance and acoustic signatures of this new generation of submarines. Both classes were fitted with Sonar 2001, which used the latest

HMS *Dreadnought* nuclear hunter killer submarine, 1963. (Naval Museum Portsmouth)

sonar processing techniques to achieve the high levels of passive performance essential for covert operations. They were also fitted with the Ship Inertial Navigation System (SINS) which provided an under-ice navigation capability that was publicly demonstrated by the surfacing of *Dreadnought* at the North Pole on 3 March1971. Not only a powerful navigation system, SINS was also a critical part of the targeting system for the Polaris missile in that it supplied geographical reference data for the launch.

Comprehensive Display System

For some time since the end of the Second World War, it had been apparent that the increase in the threat from supersonic jet aircraft was such that any attack in numbers would easily overload a manually operated action information system requiring a number of highly trained operators to maintain a warfare picture. The Comprehensive Display System (CDS) was developed in the 1950s by Ferranti to meet the new threat and it was fitted to the aircraft carrier HMS *Victorious* during her major refit, which started in October 1950 but was not completed until September 1958. CDS was an analogue computer-based display system which used capacitor banks to provide a central memory in which to store air track positional information from which target rate data could be generated. This was used to apply dead reckoning movement to display contacts using embedded code. The target data supplied to CDS came from the Type 984 air warning radar, a radar designed to detect and track up to 32 air targets simultaneously in range, bearing and height. CDS could then display this picture at operator positions throughout the ship and make it available for transmission to ships in company over an HF digital data link.

While CDS represented a significant advance in action information organisation, it was only semi-automatic as manual detection and tracking of the target was required. Although a rate aiding facility was available to update target position based on the computer's calculations, operator monitoring and intervention was always a probability in a fast moving tactical situation. There was also no automatic element of target validation or threat evaluation and these judgments still had to be made by tactical teams. Thus while some degree of data processing had now been introduced into the Fleet, a full team of radar operators was still required to input and update the

Vampire on a landing approach to HMS *Victorious*. (Naval Museum Portsmouth)

data as it was being accessed by tactical supervisors and weapon teams. Apart from the manual input requirements, CDS was also operationally limited because it was, at this stage, only capable of compiling the air picture and relied on the Type 984 air warning radar, which itself was too big to be fitted in anything smaller than a large cruiser. The delays in getting *Victorious* out of refit and rapidly advancing digital technology, plus the wider operational and ship fitting limitations, meant that CDS was effectively obsolescent when it finally came into service.

Action Data Automation

Concurrently with the production of the CDS system during the 1950s, Ferranti had been developing a first-generation digital computer, designated Poseidon. This computer comprised three full size cabinets, each with 240 removable printed wiring panels plugged into sockets that were configured into eight shelf racks. Poseidon was fully transistorised and used discrete components individually soldered into conducting tracks on epoxy resin panels to form the circuit. Connection between the panels was achieved by bringing the pins out through the back of the socket and linking the panels through complex back wiring. Connections to peripheral equipment were made through plugs and sockets at the top of the cabinets.

Functionally, the equipment comprised a program store with a maximum capacity of a now risible 16KB, a data store with a maximum capacity of 8KB and a control logic and function unit for reading the program instructions and data and carrying out the processing requirement before returning the output data to the data store. The data store used ferrite core technology which was erasable but robust in that if power supplies failed the data was retained. The clock speed was normally 500kHz with a cycle time for each computer operation ranging from 2–6 microseconds. One novel aspect of technical support built into the computer was the facility for it to carry out built-in self-checking routines and display any fault to the maintainer. This type of feature would, in due course, have a major impact on the perceived maintenance training requirements.

The first variant of ADA, DAA, used the increased power and storage capability of Poseidon to increase the total track capacity and data processing capabilities over that of CDS. This variant of the command system was installed in HMS *Eagle* during her 1959 refit and was able to coordinate all of the target information available from the ship's radar sensors, comprising the Type 984 (air warning), Type 965 (long range air search) and Type 963 (carrier approach radar). The processed information was then primarily used to provide details of contact data to labelled plan displays and Tote lists of targets and weapons available to assist with the deployment of aircraft for air defence. DAA went operational in 1965.

HMS *Eagle* post-1959 refit with DAA variant of ADA. (Naval Museum Portsmouth)

The County Class

The County Class destroyers were purpose-built to deploy the Sea Slug (GWS1) anti-aircraft missile system designed for aircraft carrier defence during the Cold War. Sea Slug was fitted with four boost motors designed to launch and accelerate it to Mach 2 within three seconds before separation of the motors and the switching in of the single sustainer motor. It was categorised as a Beam Riding Missile with missile guidance being achieved using the Type 901 radar outfit, which comprised two separate radar sets. One set was used to lock onto and track the target and the other used conical scanning techniques about an axis of rotation which was collimated with the line of sight of the tracking radar. The conical scanning movement was used to first 'gather' the missile into the guidance beam and then to measure the error angle of the missile off the target tracking beam axis. When an error was detected the missile was sent commands to manoeuvre it back towards the tracking beam so as to eventually intercept the target. The ships were designed with the magazine forward and a stern-mounted twin launcher which meant the missile, of a considerable size even without the boost motors, had to travel through the ship as it was finally prepared for firing. One advantage of the missile size was that it was judged that it could be very effective against surface targets. So much so, a GWS2 variant was produced with an anti-ship capability and this was fitted in the last four ships of the County Class, the Batch 2, or Norfolk Batch. The whole class was initially fitted with two 4.5in Mk6 gun mountings forward but, in 1974, HMS *Norfolk* had her aftermost mounting, the B mounting, removed and replaced with a trials fit of four launch platforms for shipping French manufactured Exocet (GWS50) containerised missiles to enable the deployment of the Royal Navy's first dedicated anti-ship missile system. The trials were deemed successful and GWS50 conversions were carried out on the three remaining Batch 2 ships.

The Batch 1 ships had a command system that was the DAB variant of ADA and still based on the Ferranti Poseidon computer. The main difference from DAA was that the system was able to process picture data not just in the air but also the surface and subsurface tactical environments. This capability required interfaces to all radar, EW and sonar sensors for picture compilation and to the analogue fire control computers associated with the GWS1 missile and the 4.5in Mk6 gun mountings for weapon control. The significantly increased ability to process the complete battle space picture led to the introduction of threat evaluation and weapon allocation techniques

HMS *Hampshire* operating a Wessex Mark 5 helicopter. (Collingwood Photographic Department, © MOD Crown Copyright 2010)

HMS *Kent* fitted with DAB variant of ADA and firing Sea Slug Mark 1. (Naval Museum Portsmouth)

which allowed the computer to suggest the most urgent threat – based on target parameters – and the most appropriate weapon system for engagement. Although fixed wing aircraft operations were not a design operational requirement in the County Class, the functionality for aircraft control built into DAA was incorporated into DAB and used to support helicopter deployment, particularly in anti-submarine operations using the ship's Wessex helicopter fitted with a dipping sonar and air dropped anti-submarine torpedoes.

The linking of picture information to a weapons control capability led to DAB being given the acronym ADAWS1 (Action Data Automation Weapon System Mk1). In the Batch 2 ships, the Poseidon computer was replaced by a Ferranti FM1600 digital computer and retained the same combat system functionality as its predecessor. The FM1600 still utilised discrete components but it was a step forward into the real time computation that was essential for effective combat system management. Alongside the FM1600, Ferranti also introduced a standard set of interface specifications for the associated peripheral equipments. The size of the computer installation was not much reduced at this stage from the Poseidon but the standardisation of the interfaces was a major step to more efficient computer systems design. It established Ferranti as the supplier of choice for Royal Navy computer-based command systems for the next 30 years.

An important underlying feature of the ADAWS technology was the establishment of a cartesian coordinate system, as opposed to a polar coordinate system, to denote the positioning of a target in horizontal and vertical planes, centred on the ship but fully stabilised against ship motion using gyro stabilising inputs. This was as significant a change in fire control principles as Scott's introduction of director firing as it easily permitted absolute target information to be passed to any sensor or weapon system with a positional reference linked to those true planes to use the information to acquire the target and calculate its own fire control solution. This capability allowed fully flexible allocation of contacts to sensors or weapons and could even be extended for use in naval gunfire support under indirect fire conditions.

Building on experience from the Type 81s, the County Class was also fitted with a CoSAG propulsion system but with twin screws, and this gave the ship the manoeuvrability lacking in the Type 81. Given the projected life for the class, the downside of this arrangement was that the manpower, training and support requirement was expensive as a result of the mix of technology.

The eight County Class ships came into service during the period 1962–1970 but the demise of the traditional aircraft carrier in the 1970s, following the cancellation of the CVA01 carrier programme by the Labour Government in the 1960s, meant that the prime role of the County Class was under threat and all were paid off from Royal Navy service by the mid-1980s. However, HMS *Kent*, one of the early Batch 1 ships, was retained as a harbour training ship until 1993 when she was finally scrapped.

The Leander Class

In 1963, the first of the Leander Class – at 26 hulls the most numerous class of frigates ever built for the Royal Navy – was commissioned as HMS *Leander*. *Leander* actually started out construction as a Rothesay Class but was converted during build. Initially, the Leanders were intended as general-purpose frigates but, as their longevity in service saw changes in the threat assessment, in later years many were modified to fulfil more specialist roles. Built in three batches over the period 1959–1969, the class embodied the consolidation of the technology of the time into vessels of frigate size – around 2500 tons – and a design that was economical on build costs. Such was the warship building capacity in the UK fifteen years after the war had finished that there were thirteen different shipbuilding yards involved in building Leanders over the ten years of production, including the Royal Dockyards at Devonport and Portsmouth. The only changes of significance over the ten-year period were improved boiler designs and, in the last batch of ten ships, the beam was increased from 41 to 43 feet in order to improve seakeeping and provide more internal space for future modifications. The power generation and distribution system was what had become standard for the Royal Navy – 440 volts AC 3 phase 60hz – produced by two turbo alternators with two diesel generators as backup. Power distribution was through two switchboards feeding onto a ring main system that could be divided into four sections under action conditions with normal and alternative supplies available to all important services.

Typical Leander radar outfit: Types 965, 993 and 978 surveillance radars and 903 fire control radar. (Naval Museum Portsmouth)

HMS *Achilles* standard gun Leander frigate. (Naval Museum Portsmouth)

The sensor and weapon fit changed little throughout the build programme and was epitomised in the 'Gun Leander' configuration. This configuration comprised a radar suite of the Type 978 navigation radar, Type 993 target indication radar and the Type 965 long-range air warning radar. The sonar suite comprised the Type 177 medium-range search sonar and Type 170 attack sonar. Weapons included a twin 4.5in Mk6 gun, a Sea Cat close range missile system and a single Limbo AS mortar system. Anti-submarine capability was enhanced by a torpedo-carrying Wasp helicopter, which also had an anti-ship role using the AS12 air-to-surface missile.

Apart from the high power systems technology used in the ships, which had evolved gradually since the war, the combat system equipment was largely based on discrete components employing a mix of valve and solid state technology. Ordnance engineering was very much to the fore with hydraulics being used to drive gun mountings and missile launchers as well as general services such as winches, ammunition hoists and davits.

While many weapons systems used computation to generate fire control solutions, the computers[111] were analogue and offered few interfaces to equipment other than to the weapon for which they were provided. While digital computers had made their first appearance in 1959 with CDS, this equipment was still too large for installation in a frigate-size ship. This inability to take advantage of the technology used in CDS – and subsequently ADA – meant that there was very little automation embedded in the Leander AIO system. The eponymous MATCH system was a classic example of a manually coordinated system for detecting a submarine target and finally prosecuting it using the Wasp's Mk44 torpedo. As a system it required some thirteen men to deliver the weapon and, when grossed up across a combat system, this meant that the Leander was manpower-intensive at an unsustainable level, given the pressures already being felt to reduce the running costs of the peacetime Navy.

Electronic Warfare

Dr David Kiely:

> In the early post-war days, whilst HFDF was appreciated, anti-radar electronic warfare in the United Kingdom was regarded as something of a secret oddity in the context of defence equipment and operational doctrines. Although electronic warfare had a few enthusiasts, viewed by others as rather eccentric, it was not taken very seriously, although it was agreed that on occasions it could possibly be tactically useful. It was not seen as an important element in defence equipment for sea, land or air battles as it was felt it could not be relied upon since it depended on intelligence information and enemy reaction to its use. However, this was really an expression of the lack of understanding, at the time, of the growing dominance of electronics in the conduct of warfare and the control of weapons.[112]

The latter part of the 1940s saw rapid developments in centimetric radars for both ships, aircraft and, most significantly, guided missiles. As such it became a priority to develop the technical equivalent of the HFDF set which could detect radar transmissions at centimetric frequencies, i.e. operating in the 3–9MHz range. This led to the beginning of what in the early 1950s became the specialist area of tactics, systems and equipment known as Electronic Warfare (EW).

The Royal Navy's approach to detecting and classifying radar signatures was to use wide band receivers to detect the radar transmission and a configuration of receiving horns to derive the direction from which the transmission had come. The first equipment to be deployed with this basic capability was the UA1 which was able to detect frequencies in the 3–10 cm bands[113] using amplitude comparison techniques, both visually and aurally, to detect a target and establish its direction. It was soon noted that practised operators were also able to deduce aerial rotation rates and even estimate pulse repetition frequencies from the aural signal received. This made it apparent that EW technology had not only the potential for early warning of enemy radar presence, given its range advantage over the transmitting radar, but also a greater potential for classifying the

UA1 EW outfit: 1–18GHz DF horn aerials just below the FH5 HFDF birdcage aerial. (Naval Museum Portsmouth)

nature of the transmitting platform and the specific identity of the individual platform in question. With the burgeoning threat of active homing missiles, EW capability was to become one of the most critical areas of military development. Helped by the level of operational urgency during the Cold War and the increasing interest of industry, a technique for instantaneous frequency measurement was developed by Mullard Laboratories that enabled the frequency of each intercepted pulse to be measured. With this core capability in place, equipment design moved swiftly forward to included more comprehensive signal analysis which could determine not only the frequency of each transmitted pulse but also pulse repetition frequencies and aerial rotation rates.

The transistor made an early contribution to EW when it was used in the UA4, a solid state version of the earlier UA1, and this first enabled EW installations in submarines. Here the equipment was used when submarines were snorting on the surface to give early warning of any aircraft radar in the vicinity. By the 1960s, solid state technology was being used in the UA8, 9 and 10 equipments to compare received signals against a library of known radar 'fingerprints' from which equipment sources and platform information could be deduced. Equipments which were capable of using received radar transmissions for the detection and classification of hostile platforms by own ship became collectively known as Electronic Support Measures (ESM).

The ESM designation was to differentiate such passive operating equipment from equipment which evolved out of the next technology development: Electronic Countermeasures (ECM) technology. ECM was used for the disruption of hostile transmissions with the aim of confusing any information gathered by the enemy platform for tactical advantage, including the homing phase of an approaching anti-ship missile. Early applications involved the use of noise jamming across the band in which the hostile equipment was radiating. This was quickly shown to have a downside as the enemy's own ESM systems were equally able to establish own Force position based on the source of the jamming noise. This fundamental dilemma in the tactical use of jammers was to be the subject of future debate for, even with the development of more sophisticated electronic jamming and decoying techniques, many continued to feel that electronic silence was the most reassuring tactical stance.

ECM was also applied to the technology used in off-board decoys, most of which, at the time, were passive in operation relying on the threat radar being distracted by the radar returns from the decoy. These early decoys were proven to be effective as most radars could not differentiate between ship and decoy returns. This type of weakness in radar performance led to the introduction of the final facet of EW, Electronic Counter-countermeasures (ECCM). ECCM involved prevention of the enemy disrupting the transmitting ship's radar transmissions by using electronic techniques, such as frequency hopping, which made successful ECM interference by the enemy more difficult. At this time ECCM technology was very much in its infancy. The main development objective was to keep ahead of any improvements to ECM capability.

Such was the speed of advances in EW technology and the secrecy surrounding it, it became a specialist area of knowledge for both users and maintainers, with a consequent effect on branch structures.

CHAPTER 10

THE WEAPONS AND RADIO BRANCH

The Engineering Specialisation

In 1960, the Engineering Specialisation Working Party was tasked to look at the continued suitability of having an Electrical Branch and an Engineering (viz. mechanical) Branch in a technological age when the systems coming into service with the County Class destroyers – particularly the Sea Slug missile system and the gas turbine propulsion system – required a broader enineering knowledge that was actually spread across the two existing technical branches and the Ordnance Branch. The investigation concluded that the time had come for technical officers to break away from the segregation of professional disciplines and take on responsibility for complete systems. It was the ship staff experience of maintaining the Sea Slug system during its development phase in the trials ship, HMS *Girdle Ness*, that gave momentum to the drive to merge ordnance and electrical expertise into one branch and raised the issue of the future weapons role of the electrical officer. It was found that the close integration of weapon, fire control system, launcher and magazine embraced many technical disciplines and, in conjunction with the sheer size of the system, required very close liaison not only between the 'O' and 'L' departments but also the rest of the ship. The recommendation from *Girdle Ness* was that a combined department was the most efficient way of achieving the system maintenance task and ensuring sufficient priority for the whole ship inputs needed to support that task.

With a greater breadth of training in mind, the ESWP confirmed the Carlill recommendations that there should be one 'engineering specialisation' with three engineering branches inside the

(Collingwood Magazine Archives)

specialisation. The three branches were to be the Weapon and Radio Engineering (WRE) Branch, the Marine Engineering (ME) Branch and the Air Engineering (AE) Branch. This decision was predicated on the Ordnance Branch being subsumed into the WRE Branch. With this concept of an engineering specialisation in place, it was considered that the surface ship could retain two engineering departments while the Fleet Air Arm squadrons and the submarines could have an engineering specialist in overall administrative charge of a department which comprised two specialist mechanical and electrical sub-departments.

It is important to note that the substance of this change of structure was really only applicable to GL technical officers. Effectively it altered the scope of the training they would need in order to undertake a wider breadth of technical management. The change did not materially affect the electrical SD List officer, except to produce a change in title whereby the 'electrical' prefix, awarded in 1946, was replaced by the generic prefix 'engineering'. The specialist suffices were retained and extended to include the OE specialist. Thus the new SD ranks became, for example, Engineering Lieutenant (L), (RE) or (OE).

The Electrical Branch rating structure was also largely unaffected by the change as the line management remained unaltered. The problem was the future ordnance engineering requirement, for despite the concern that there might be a skills gap in available ordnance expertise, the decision was taken to cease the training of OAs with the last apprentice class due to leave *Caledonia* in 1964. This gap might have been filled by armourer ratings from the gunnery and TAS branches but the ESWP considered that it was essential that any specialist ordnance requirement should be met from within the WRE Branch. With this in mind, the Weapon Mechanician (WM) rate was introduced as part of the branch structure.

The Weapon Mechanician

The WM was generally to be drawn from seaman armourers and, in common with the existing mechanicians in the ME and WRE Branches, the WM candidate undertook a two-year course at *Caledonia* with the first course starting in January 1962. The first candidates were mainly petty officers and leading ratings with G or TAS qualifications but as these numbers dwindled, probably owing to the technical experience of the TAS and Gunnery Branches reducing because of the creation of the WRE Branch, so the candidates became predominantly able ratings who were less well equipped in terms of experience for the WM course. With this downward trend identified, the direct entry Weapon Mechanician Apprentice was introduced in order to sustain the specialisation.

Chief Naval Engineering Officer

On 1 July 1961 the concept of an overarching engineering specialisation was officially established by the introduction of the post of Chief Naval Engineering Officer (CNEO) which was taken up by Admiral R. T. Sandars, a mechanical engineer by profession. CNEO was given the responsibility within the Department of the Second Sea Lord for all engineering personnel, giving them a single voice in the Admiralty Board. In recognition of the fact that certain engineering appointments were to be filled by any engineering specialisation, the Directorate of Naval Officers Appointing (Engineering) (DNOA(E)) was established. This ensured the permeation of multi-engineering discipline experience up through the hierarchy and an informed and objective view of the impact of technological progress on the shape of the Royal Navy. Undoubtedly, this was an invaluable facet which underpinned Engineering Branch development for the future. In 1967, Rear Admiral Hughes became the first electrical officer to be appointed to the post of CNEO.

A Systems Management Approach

The boundaries of responsibility between the new engineering branches were announced in AFO1342/61 and these were defined by system function and not technical discipline. Under these arrangements the material responsibility for a primary system, regardless of the technology involved, was to be clearly with a single technical department, even though the specialist rating support needed may have been embedded in another department. It was anticipated that this need to borrow expertise would be continued until rating training had evolved to allow the responsible department to undertake its own support duties.

With this principle of ownership established, the major impact on the electrical officer was subtly to change the arrangements established in 1946 whereby the user officers had officially retained responsibility for their equipment, releasing it to the electrical department for maintenance. Now, under the terms of AFO 1342/61, the Electrical Branch was to be formally given responsibility for the systems it maintained, namely all electrical (including radio and radar) and weapons systems. Since weapon systems closely embodied both electrical and ordnance technology, it was considered necessary to transfer ordnance engineer officers and ratings into what was then designated as the WRE Branch.

In order to prepare serving officers for what was a new technical responsibility, in March 1961 *Collingwood* started electrical cross training courses for marine engineers and ordnance engineers. The former required it for seagoing duties, which included oversight of electrical matters, and the latter needed it in order to be re-categorised as Weapons and Radio Engineer Officers and be eligible for, primarily, seagoing posts. Under this arrangement the re-categorisation of the ordnance engineers from the Ordnance Branch to the WRE Branch was geared to the appointing cycle and in some cases this meant that some officers would never become sufficiently cross-trained to qualify for WREO status. A further complication was that the Ordnance Branch officer structure was shrinking rapidly and this meant that OA's promotion prospects were being severely limited for the future.

GL electrical officers were not deemed to require any formal cross training to be re-categorised as WREOs because their existing training profile contained sufficient technical breadth. However, in line with the Carlill Committee recommendations, from September 1960 all new entry GL engineer officers entering the Royal Naval Engineering College Manadon undertook a common degree course training, which gave them the breadth of electro-mechanical knowledge needed to fulfil the role of either a WREO or an MEO.

(Collingwood Magazine Archives)

CHAPTER 11

THE INTEGRATED COMBAT SYSTEM 1967–1978

During this period the Navy continued to shrink from a manpower total of 100,000 in 1967 to 76,500 in 1978. The government withdrew from permanent commitments east of Suez and the Far East Fleet was abolished in 1971. The replacement carrier programme had been cancelled and naval platforms of cruiser designation and above were declared obsolescent. In 1972, the Principal Warfare Officer (PWO) was introduced and the Operations Branch was formed in 1975 as the new Executive Branch charged with fighting the ships.

The Type 82 destroyer

The Ferranti Poseidon computer-based system, because of its size and ship's services requirement, had been limited to installation in larger ships. However, the options for fitting a similar command system capability into much smaller ships had increased significantly with the development of the FM1600 computer, first used in the Batch 2 County Class. The time of a fully integrated sensor picture as well as direction and fire control of the weapon systems to form a fully integrated combat system was approaching and it arrived in the form of HMS *Bristol*, the only Type 82 destroyer ever built.

HMS *Bristol* ADAWS2 fitted Type 82 destroyer. (Naval Museum Portsmouth)

In 1973, *Bristol* was commissioned with a combat system conceived around the real-time capability of the FM1600 computer. Initially, the ship was designed as an area defence platform for the now defunct CVA01 aircraft carrier. The ship was not much smaller than its County Class predecessor – 6000 tons versus 6800 tons – and it became the test bed for a number of new systems. Some of these systems reflected major advances in technology, such as the FM1600, Sea Dart (GWS30) and the 4.5in Mk8 gun with its associated GSA1 fire control system. However, many other systems used valve technology, including the Mortar Mk10 and its associated fire control sonar Type 170, the medium range sonar Type 184 and the Ikara missile-carrying torpedo system (GWS40), which still required high-level use of auxiliary services and, most importantly, a significant number of men for both maintenance and operation. On the manpower front, it is also worth noting that the GWS30, even though at the cutting edge of technology, still required nine men for its operation at action stations, not including the men needed for the picture compilation, which by now had achieved a degree of automation, and the command function. Many of these GWS30 operators were in fact WRE Branch maintainers and were shown as user-maintainers in the ship's watch and quarter bill.

The command system in *Bristol* was designated ADAWS2 and it became the leading example of an integrated combat system in the Royal Navy. Unfortunately, the real benefits of widespread use of solid state electronics and automation had not been achieved and there had there been no significant reductions in two of the critical cost drivers of the time, namely manpower and training. Perversely, from an electrical technology point of view, the wider use of solid state, printed electronic card (PEC) and digital technology alongside the valve had broadened the scope of technical training needed for both the electrical and radio specialisations. The consequence of this was that the shore training bill was going up and the existing WRE Branch structure was already starting to look inadequate.

In the Marine Engineering Branch, manpower issues were also aggravated by the fact that the propulsion system was modelled on the County Class and used the CoSAG configuration of steam turbines for cruising and Olympus gas turbines for high speed manoeuvres. This technology mix doubled the shore training bill, retained the on-board manpower needed for boiler room watchkeeping duties and required the MEO to have control of electrical expertise (for machinery control systems), which had been vested in the Electrical Branch and its successors since 1946.

The Type 42 Destroyer

After the procurement programme for Type 82 air defence cruisers was terminated, the Type 42 was proposed as a lower cost option, smaller by some 1000 tons but, because of the use of modern technology, with essentially the same capabilities as the Type 82. Designated as the Sheffield Class and classified as an air defence destroyer, the Type 42 retained many of the newer systems featured

HMS *Newcastle* firing GWS30 Sea Dart. (Private Collection)

in its prototype forerunner. The GWS30 and Mk8 gun were fitted for air and surface operations but the GWS40 and MCS10 anti-submarine systems were replaced by STWS1 (Ship Launched Torpedo Weapon System 1) and a flight deck and hangar were added to allow the deployment of the Lynx helicopter in anti-submarine and anti-ship roles. The command system was the ADAWS4 variant, which used twin Ferranti FM1600 computer-based systems, one for picture compilation and the other for weapons control.

Type 42 ADAWS4 operations room. (BAE Systems Archives)

HMS *Newcastle* ADAWS4 fitted Type 42 destroyer in the turn. (Private Collection)

One radical change from *Bristol* was that the Type 42 propulsion comprised a Combined Gas or Gas (CoGOG) system with two Tyne gas turbines installed, instead of steam turbines, as the main cruising engines and the two Olympus engines being retained for high speed manoeuvring. Effectively this removed steam technology from this class of ships and, given that fourteen ships were eventually built, this generated significant space, manpower and training savings. Moreover, these changes signified a major change in the scope of technical expertise required by marine engineering departments in Type 42 vessels, with the emphasis moving rapidly to electronic machinery control systems, and support of the major propulsion units becoming repair by replacement rather than craft-based repair.

Computer Assisted Action Information System

In May 1974, the first of the Type 21 frigates, HMS *Amazon*, entered service fitted with another Ferranti FM1600 digital computer-based system, the Computer Assisted Action Information System (CAAIS). CAAIS was based on the FM1600B computer which used integrated circuit technology and utilised the standard interfaces to the combat system equipment previously developed by Ferranti. Although functionally similar to the FM1600, the FM1600B was physically smaller and this enabled CAAIS to be produced in a number of DBA-designated variants for

HMS *Amazon* CAAIS fitted Type 21 frigate. (Naval Museum Portsmouth)

CAAIS Command System operations room. (BAE Systems Archives)

fitting in both large and small ships and designed to fulfil a variety of operational roles. Following on from the Type 21, anomalously fitted with DBA(2), came a range of Leander role conversions with the first being DBA(1) in HMS *Cleopatra*, commissioned in November 1975 as the first of six Batch 2 Exocet-converted Leanders. Subsequently, five Batch 3 Sea Wolf converted ships were fitted with DBA(5), a system based on the much smaller FM1600E computer and the next generation in the FM1600 computer series. Other platforms fitted with CAAIS included HMS *Hermes* (DBA(3)), the *Hunt* Class of mine countermeasures vessels (DBA(4)) and, in due course, the first four Type 22 frigates, which were fitted with DBA(5).

CAAIS did not have the same track management capacity or integrated weapons control functionality of ADAWS. Fire control functions were generally carried out by a separate computer, which in the case of the Type 21 WSA4 gunnery system was based on another FM1600B. Other stand-alone weapons systems linked to CAAIS included GWS25 Sea Wolf and Exocet. From an electrical technology point of view, in both the propulsion and combat system areas the Type 21 was on a par with the Type 42 and reinforced the need for the Royal Navy to review its branch structures and training programmes.

The Leander Conversions

During the 1970s, one problem with carrying out any structural review of engineering support in the Royal Navy was that the life of the Leander was being extended as advantage was taken of the computer power now being made available to introduce combat system upgrades. The earliest upgrade programme involved the installation of a single FM1600-based ADAWS system to support the deployment of Ikara in eight ships of the class. This was followed later by a programme to fit CAAIS in support of the deployment of Exocet and, subsequently, Sea Wolf, the latter starting with HMS *Penelope* as the GWS25 trials ship in 1972.

It was not just the surface systems which were benefiting from new technology. Initially the Leanders were fitted with Type 177 and Type 170 sonar but under the Exocet and Seawolf programmes these were removed and later ships were fitted with the new generation Type 184 medium range search sonar which used solid state electronics. At the same time the Mortar Mark 10 mountings were also removed, reflecting the reduced operational requirement for a dedicated close-range anti-submarine capability. This not only reduced the equipment footprint making compartment and deck space available for other uses, it also reduced the display requirements and number of operators in both the sensor and weapons areas. Space savings meant that in addition to its new-generation hull-mounted sonars, a number of the Leanders could be fitted with the Type 199 variable depth sonar and, in the early 1980s, the Type 2031 towed array sonar, both first fits of their sonar type in the Surface Fleet.

HMS *Cleopatra* CAAIS fitted Exocet Leander. (Naval Museum Portsmouth)

Existing equipment was also being re-engineered using solid state technology to improve availability, reliability and maintainability (ARM) characteristics and installation space envelopes. This type of equipment upgrade allowed earlier classes of ship, such as the Rothesays, to draw some benefit from technological advances. A typical example of such a programme was the upgrade of the Type 170 attack sonar used with the Mortar Mk10, which, in 1973, was re-engineered using modular component technology. This change reduced the number of equipment cabinets from three to one, eased the demands on the ship service systems and reduced the maintenance load through the use of more built-in test facilities.

Unfortunately, under the Leander upgrades, it was never a realistic option to change the main propulsion systems from steam turbines. This created a technical anchor by preventing any major reductions in marine engineering training in the steam technology area while gas turbine training was on the increase. All this came at a time when training times and manpower costs started to come under considerable scrutiny.

Communication Systems

In 1944, it was internationally agreed that Armed Forces tactical communications should be conducted in the 200–400MHz band and this took it into the Ultra High Frequency (UHF) band.[114] In addition to improved security due to the atmospheric limitations on transmission range, UHF offered more effective line-of-sight communications with aircraft. This decision led to the development of the Type 691 UHF transmitter/receiver, which was starting to be fitted in the Fleet by 1953.

Although the concept of a whip aerial had been introduced by the US during the war and considered by the Royal Navy in 1944, it was not until around 1954 that it was introduced into the Fleet. The advantage of the whip aerial was that it did not require any top support and could therefore be installed in areas clear of other electrical interference sources. Although heavy, its rugged nature meant that it was less prone to battle and weather damage. The fact that the aerial could be sited remotely from the communications office meant that it was necessary to mount the 30-foot aerial on a base tuner unit. The base unit could then be remotely tuned such that the whole whip aerial assembly with its feeder appeared, effectively, as an electrical length of a quarter, half or full wavelength at the operating frequency. This matching technique reduced the power being reflected back into the transmitter and optimised power transfer into the aerial for maximum transmitted power output. Initially, this type of aerial was fitted in carriers as it could easily be lowered into a horizontal position during flying operations. The whip aerial was used mainly for HF transmissions and its use was quickly extended to smaller classes of ship.

HMS *Grenville* SCOT satellite communications trials ship *c*.1970. (Naval Museum Portsmouth)

(Collingwood Magazine Archives)

Communications crossed another technology threshold in November 1969 when the Skynet 1 satellite communications system came online. The geo-stationary satellite was placed in orbit above the Indian Ocean and it relayed communications services between eight earth-based stations. Ashore there were facilities at Oakhanger in England – which acted as the controlling base – Hong Kong, Gan and Cyprus. Skynet was a tri-Service system and there were two Royal Navy systems, designated Type 5 terminals, initially installed in HMS *Fearless* and HMS *Intrepid* during the late 1960s and later transferred to HMS *Ark Royal* and HMS *Hermes* in 1971. The performance of the Skynet Type 5 terminals was so successful that, in 1970, the Type 15 frigate, HMS *Grenville*, was used to carry out trials with a new Satellite Communications Onboard Terminal (SCOT), a transportable communications cabin which could quickly be installed in destroyers and frigates. Speed of installation was achieved by providing selected ships with a 'fit to receive' capability which permitted the cabins to be moved around the Fleet to support the roulement style of operations that were becoming a pattern for the Royal Navy. SCOT was a dual antenna system and the first terminal went to sea in HMS *Blake* in 1974.

Skynet 1 and its successor systems took Royal Navy communications into the Super High Frequency (SHF) band and higher, and the transportable SCOT facility meant that ships down to destroyer and frigate size could be made capable of maintaining worldwide secure voice and teletype communications prior to deployment.

A Shift in the Balance

By the end of the 1970s, nearly all major surface vessels had, or were due to be fitted with, modern missile and gun systems, digital computer-based operations rooms, more efficient and reliable sensors and satellite communications facilities. A significant number of ships also had some gas turbine propulsion facility with the increasing number of Type 42 destroyers, Type 21 and Type 22 frigates being reliant totally on the gas turbine. The Royal Navy by now was well into a new technology era and the impact on trained manpower and branch structures was, once again, starting to be of concern.

CHAPTER 12

THE WEAPONS AND ELECTRICAL ENGINEERING BRANCH

The Engineering Specialisation Working Party

As a result of recommendations of the Engineering Specialisation Working Party in 1964, in DCI(U)416/65 of March 1965 the Admiralty Board stated that there was a need to introduce a viable structure to replace the WRE Branch and incorporate the remaining personnel of the Ordnance Branch who had not yet been cross-trained and transferred into the WRE Branch. The Admiralty wanted to ensure that the foreseeable Royal Navy requirement for electrical engineering support would be met and at the same time they wanted to put into place a structure that would provide equitable avenues of advancement and promotion for all officers and ratings in the electrical engineering community, including the ordnance fraternity. This meant allocating the engineering tasks to give each man an equal measure of both responsibility and opportunity. Accordingly, they approved in principle the introduction of the Weapon and Electrical Engineering (WEE) Branch. This structure was to be comprised of GL and SD weapon and electrical engineering officers supported by radio, control and ordnance specialist categories of rating.

The New WEE Branch Structure

By 1966, most GL ordnance engineer officers had been cross-trained to become WREOs and it was then that ordnance engineer officers on the SD List were given a change of title that brought them into alignment with the new WEE Branch. They already had the prefix 'Engineer' in their title, and when they transferred to the WEE Branch they were given the suffix (OE) alongside their (RE) and (L) specialist counterparts.

The three WEE Branch rating specialist SQ streams were also introduced in 1966 with each stream having its substantive rate structure and containing artificer, mechanician and mechanic non-substantive qualifications. The three artificer specialists were categorised on 1st February 1967 as Radio Electrical Artificer (ex-REAs), Control Electrical Artificer (ex-CA(W)s and selected OAs) and Ordnance Electrical Artificer (ex-EAs and OAs). The re-categorised OAs retained a suffix (O) in their CEA or OEA until they had been sufficiently cross trained. Where, previously, selection for officer had been largely limited to those with artificer qualifications, it was now opened to mechanicians. Thus the promotion avenue from mechanic to admiral was opened and selection for mechanician training was a prize sought by many mechanics in the subsequent years because of the opportunities it offered.

Although there was some redistribution of responsibilities, the tasks of the RE and CE streams corresponded broadly to the earlier radio and control tasks. However, the OE stream combined the Ordnance Branch weapons mechanical element with the heavy electrical power element of the old Electrical Branch. Thus, at the semi-skilled level, the Ordnance Electrical Mechanic's (OEM) duties were a combination of part of the EM's duties and the semi-skilled maintenance work carried out by seaman armourers. At a stroke this meant that the OE structure was supportable from new entry mechanic rating to officer and the weapon mechanician scheme put into place to fill the perceived ordnance skills gap was no longer needed.

Where necessary, ship complements were altered to reflect the new structure. The billets mainly affected were those filled by the L and OE specialisation where the combined skills of heavy electrical, ordnance and hydraulics were requirements for holding each newly defined billet. Subsequently, DCI(RN)1235/70 decreed that any L, OE, TAS or G rating who had completed twelve months in a technical billet at sea affected by the change, and had satisfactorily performed the combined duties for six months, could put up the new badge and suffix for his WEE Branch rate.

Collingwood was reorganised to form three specialist category training schools (control electrical, ordnance electrical and radio electrical) and, following new entry training at HMS *Raleigh*, all WEE ratings graduated from one of these schools. In support of the new responsibility for hydraulic machinery, a hydraulics training capability was installed in order to cope with the requirement for gun mounting maintenance. With basic hydraulic training in place, although most of the gun mountings were left on site at the gunnery school in *Excellent*, *Collingwood* was then proclaimed as the Royal Navy's Weapons and Electrical Engineering Training Establishment. Experience of the three-pronged structure over a number of years showed some overlap in the skills needed by the control category with those needed by the ordnance and radio categories and this resulted in some subject matter during the early stages of training being common to both.

Shortly after the introduction of the Fleet Chief, the decision was made to combine the artificer apprentices and mechanician apprentice courses. DCI(RN)1140/71 laid down that from 1 January 1972 the mechanician apprentice scheme would be abolished and the conditions appertaining to the artificer apprentice entry were changed to allow for the direct intake of suitably recommended Mechanics First Class. As part of the change the qualifying age for entry to the artificer scheme was extended to accept mechanics from fifteen years six months to 21 in order to catch the late developers.

Fleet Chief Petty Officer

For some time the Royal Navy had considered the establishment of a rate which would be the equivalent of the highest warrant officer in the other two Services. Apart from being able to give such a rate special status and privileges, it was felt that it would also lay to rest some of the rivalry that arose when, in a tri-Service situation, relative rates and ranks could become an issue, such as when military command needed to be exercised or in a mess environment where mess presidencies were conferred by seniority. This lengthy gestation period ended with the introduction of the Fleet Chief Petty Officer in 1970. While the Fleet Chief equated to the highest grade Army or RAF warrant officer, it was not intended to be equivalent to the Royal Navy warrant rank which had been abolished in 1949. Instead it was to represent a further promotion opportunity for chief artificers who had not been selected for the SD List. The Fleet Chief messed with chief petty officers and wore a distinctive cap badge and insignia on the cuff. One privilege afforded to the Fleet Chief was an entitlement to be called 'Sir' by all rates junior to him.

Technology Marches On

On the equipment front, technology had progressed from valves through solid state and into silicon chip technology and associated computer-based systems, all in a matter of ten years. The ME department found itself responsible for gas turbine propulsion machinery with complex electronic control systems, while the operations department had to deal with a rapidly increasing mass of electronically produced information on weapon system performance requiring compilation, analysis, evaluation and rapid executive responses. These trends started to place new demands upon the WEE department with the WEEO assuming responsibility for a breadth of tasks ranging from maintainer support for propulsion control systems closely integrated with the operations of the ME department to quarters training of WEE personnel manning combat system peripherals such as the Type 909 radars, Sea Dart Launcher and the 4.5in Mk8 gun in the Type 42s.

A further concern was the perceived dilution of explosive ordnance expertise in the Fleet with the subsuming of the Ordnance Branch into the WEE Branch. This was perhaps as much due to the natural evolution of the naval gun as it was to the new WEE maintainer system. In earlier capital ships the big gun installations had penetrated many decks and formed large systems which required a dedicated Ordnance Department, led by ordnance officers and employing many OAs, who rose up through the branch structure as their experience increased. Always there were more senior men in the gun to turn to for advice and assistance. With the demise of the capital ship, the more modern ships were only fitted with a single medium calibre gun turret (4.5in Mk8) complemented with a single ordnance electrical artificer. Almost invariably that artificer joined straight from his apprenticeship to take charge of a highly sophisticated turret and magazine which, although much smaller, nevertheless retained all the explosive and hydraulic hazard elements of the ordnance systems of the past.

In 1972, the CNEO, Vice Admiral Sir George Raper, wrote a paper in which he drew attention to the increasing growth of combined electro-mechanical systems, particularly those associated with gas turbine propulsion plants in surface ships. He also stated the fact that sooner or later ME departments would have to be able to deal with their own electrics, a position reached some years earlier in nuclear submarines. He also observed the increasing demands that the principal warfare officer and user/maintainer systems were making on the WEE department's time and thought it desirable to reduce the range of the WEEO's responsibilities. Accordingly, the CNEO proposed another deep study of the whole engineering specialisation to look at the reallocation of technical responsibilities, since the problems clearly crossed all sub-specialist boundaries, and to consider the impact of the WEEO's warfare commitments on his technical role. The Admiralty Board approved the conduct of this study and the Engineering Branch Development Study was announced in DCI (T)(RN)126/73.

Vice Admiral Sir George Raper, the Chief Naval Engineer Officer. (Collingwood Museum)

CHAPTER 13

THE WEAPONS ENGINEERING SUB BRANCH

Engineering Branch Development Study 1973–74

The Engineering Branch Development (EBD) study, as recommended by the CNEO, was given to an in-house Engineering Branch Working Group (EBWG). The group was tasked to consider a time frame circa 1985 and given the following terms of reference:

> In the light of developments in engineering and weapon systems in the Royal Navy and the likely availability of resources, to study and recommend how the structure and training of officers and ratings of the Engineering specialisation and its sub-specialisations should be developed so as to provide for effective operation, support, development and design.
>
> The study was to be thorough and detailed, taking two years in its deliberation and involving, to varying levels, the examination of every aspect of the Ministry of Defence (Navy). Its findings were to be comprehensive, far sighted and far reaching and, since both the ME Branch and the WEE Branch appeared to be working well, any proposed changes were to be made in a controlled and systematic manner over the next decade. The fundamental policy underpinning the study was the requirement to adopt a systems approach in all engineering fields and the need to adopt a user/maintainer concept where appropriate.

There were three drivers that fuelled the need for new branch structures and training programmes, and they were all associated with advancing technology.

Expansion of Electrical Technology

One of the drivers for change was the belief that the steady expansion of electrical and electronic technology into every aspect of naval engineering made it no longer practicable to have non-indigenous electrical engineering expertise outside the control of, particularly, the ME Branch. It was a testament to the speed of technology advances that this conclusion was reached less than 30 years after it had been determined that consolidation into the Electrical Branch was the answer to coping with the perceived progress of electrical technology.

The circumstances which had led to a change in view are broadly covered in this book. However, one possible single explanation might be the appearance of closely integrated systems using complex microelectronics, particularly in the ME Department. These types of system had few clear technical boundaries along which departmental

(Collingwood Magazine Archives)

responsibilities based on specialist expertise could be aligned. Arising from this, two of the questions to be answered by any review were whether any clear boundaries did in fact exist and, if they did, how long would they be in existence given that technology was still moving forward.

WEE Branch Warfare Responsibilities

The second motivation was the growing involvement of WEE officers and ratings in the operational use of weapons and sensors. This increased involvement was by now a fact of life, with most weapon quarters personnel in modern ships being made up of WEE ratings. For example, in the Type 42 there were four senior ratings at the Type 909 desks, two senior ratings in the Sea Dart magazine, one senior rating as Captain of the Gun and another in charge of the torpedo preparation party. All of these quarters positions were supported by WEE junior ratings or seamen on loan to the WEE department for duty and classed as WEE Seamen.[115] In the operations room one of the WEE section officers was also designated in the ship's quarter bill as a Defence Watch Sea Dart Controller.

In addition, the recording and analysis facilities built into the command systems provided a source of both weapon system and operator performance data. As owner of the equipment, the responsibility for extracting and analysing this data fell to the WEEO with the outputs being, amongst other things, performance assessment of the Operations Branch personnel. This capability was something in which the PWO would take great interest and seek detailed input from the WEEO.

Professional Institutions and Qualification Boards

The third driver was that the Royal Navy had a vested interest in keeping its uniformed personnel professionally recognised by its own civil service professional engineering and technical staff. Also with an eye to recruitment, it was obliged to ensure that its training and branch structure aligned sufficiently with the technical scope of the national institutions to ensure personnel could be accredited to those institutions in terms of both qualification and experience. In this context, the civilian engineering institutions were at the same time adapting their accreditation criteria and technical scope in order to ensure the continued credibility of their own discipline-based engineering professions and to keep up with evolving technology and systems with similar complexity to those in naval service. Consequently, these national institutions were looking for a similar technical breadth in Royal Navy candidates seeking professional recognition as engineers at the national level. With professional recognition in mind, a key aspect of the Royal Navy's training policy for both officers and ratings at the time was that any course design should be cognisant of any related civilian qualification. A rule of thumb was that if a civilian accreditation could be achieved with no more than 10 per cent increase to the training time, then this was to be accepted.

(Collingwood Magazine Archives)

EBD Study Recommendations

The EBWG examined the Board guidance and identified a number of key areas of interest for investigation in what was to be a closely coordinated study programme. The report was published on 31 March 1975 and made the following major recommendations:

> All Engineering functions should be brought together under one Engineering Branch with Weapons Engineering (WE), Marine Engineering (ME), and Air Engineering (AE) as three sub branches within the Engineering Branch.
>
> The ME Sub Branch should be developed over the next decade into an integrated electro-mechanical organisation responsible for the hull and its systems, propulsion systems and power generation and distribution systems including associated electronic elements.

The case for transferring some electrical responsibilities to an ME Sub Branch was found to be a strong one as the use of gas turbine propulsion was by now extensive in the Fleet, with all major warship classes in build planned to have such propulsion. Although it was not known at the time, the Type 23 frigate – designed in the late 1980s – was to add to the case as it was to be fitted with a diesel electric propulsion system, thus increasing the breadth of electrical expertise needed by the ME Sub Branch.

The EBWG was also concerned to note that an unacceptable decline in standards associated with the custody of ordnance had become evident due to the confusion over weapons equipment responsibilities as a result of the development of the Operations Branch.[116] In view of this deterioration, it was considered that the primary task of the proposed WE Sub Branch over the coming decade was to re-establish the high professional competence and sense of commitment associated with weapons and ordnance engineering. The efficiency of the end product, the Fleet and personnel safety were all considered to be directly involved. Closely allied to this was the need to develop a greater depth and continuity of experience in those who were to maintain the complex weapon systems for the future. Discussions with other navies about an efficient Operations/WE departmental interface had revealed that the problem of optimising the management of modern weapon systems was not peculiar to the Royal Navy, but that it was generic to developments in automated weapon and sensor systems, which were making the roles of user and maintainer less distinguishable. The French and Canadian Navies had initially adopted a total user/maintainer policy for their weapon systems but found that the standards of upkeep seriously declined. Subsequently they had partially reverted back to their old structures.

In order to achieve more efficient management of weapons systems, afloat and ashore, and to make better use of the skills and experience available, the EBWG recommended that a detailed study should be carried out with Sub-Navy Board authority to recommend the best method of establishing a closer relationship between the Operations Branch and the new WE Sub Branch, as it was evident from their research that WE officers, particularly in frigates, were being overloaded by their increasing involvement with operational and administrative tasks.

GENERAL SERVICE JUNIOR WEAPON ENGINEER OFFICERS' APPLICATION TRAINING AT HMS *COLLINGWOOD*

HMS COLLINGWOOD

(Collingwood Magazine Archives)

The report also expressed concern that, in the 1980s and beyond, there would be difficulty in meeting recruiting targets due to the demographic situation. The introduction of the notice engagement was also likely to make it difficult to get the continuity in service which had been the norm under the Long Service and Reserve (LSR) and Continuous Service (CS) schemes. The Navy Board agreed in principle with the final EBD Report and gave approval to the establishment of a new Engineering Branch with the proposed three specialist engineering Sub Branches, WE, ME and AE. The AE Sub Branch was already an engineering entity in its own right and so it was the realignment of the ME and WE electrical responsibilities that was going to be the challenge.

EBD Implementation

Under EBD, the MEO was to be given complete electrical responsibility for his department and three new categories of ME Artificer were introduced. These were known as Marine Engineering Artificers with the specialist suffices of (ML), (EL) or (H). The ML was to be the mechanical specialist with a complementary electrical knowledge, the EL was to be the electrical specialist with a complementary mechanical knowledge and the H was to be a re-categorisation of the previous Shipwright Artificer. The ML and EL artificers were to be supported by marine engineering mechanics who would have the specialist suffix of (M) or (L). The ME department on board ship was then to assume responsibility for power generation and distribution, ship's services and domestics as well as the propulsion control systems for the gas turbine units.

In recognition of an initial need for some WE officer expertise, a number (mainly SD(L) officers) were to be transferred and given the title ME(L). These officers were, in general, SD officers with an electrical background and they were unlikely to go to sea as they did not have any ME Charge qualifications. In order to fulfil the MEO's electrical duties on board ship, as an interim measure, rating personnel with the appropriate electrical expertise – mostly OEAs and OEMs – were to be transferred into the ME Sub Branch. They were to be given the titles Marine Engineering Artificer MEA(L), for the artificers and mechanicians, and Marine Engineering Mechanic MEM(L), for the mechanics, with continuity being eventually being achieved with MEA (El)s and MEM(L)s coming through the training pipeline. As in the case of the ME(L) officer, these ratings had no ME watchkeeping qualifications and were not, under the terms of their transfer, obliged to gain such qualifications. However, the nature of their duties allowed them to go to sea and some did gain ME qualifications.

The implementation of the change was started in 1979 and carried out on a ship-by-ship basis. Each ship's scheme of complement and quarter bill was examined to establish which billets should be transferred and what the final scope of those billets should be to ensure that no equipment was unaccounted for on completion of the transfer of personnel. Once all transferring billets had been identified, the seagoing incumbents were earmarked for the ME Sub Branch and formed the initial list of transferees from the WE Sub Branch. Each identified ship electrical billet to be transferred was given dual qualification as either the new ME Sub Branch title or the appropriate transferee title. A day was selected in the ship's programme for the transfer of responsibility to take place and on that day the ship was audited on its readiness to change before a formal handover report was signed and sent to C-in-C Fleet for the record and to initiate transfer of the ship's administrative authority arrangements.

The transfer throughout the Fleet was completed in 1981 and by that time 2,600 WE ratings had been transferred from the WE Sub Branch to the ME Sub Branch. WE transferees continued to serve their time in the new transfer categories and subsequently they were drafted to only the dual billets identified in the ship's scheme of complement. Over the next few years, power generation and distribution training was gradually transferred from *Collingwood* to HMS *Sultan*, the Marine Engineering Training School. This move resulted in the closure of a well known post-war electrical landmark in *Collingwood*, the White City Heavy Electrics Training Section.

The reduction in the scope of the responsibility for the WEO was considered to be such that he could continue with the oversight of his quarters training responsibilities. In association with

NOW I'LL GIVE
YOU SOME TIPS ON
HOW AMMUNITION
SHOULD BE
HANDLED

(Collingwood Magazine
Archives)

this, the WEO took on responsibility for explosives accounting and magazine safety from the
Operations Branch although no transfer of staff accompanied the transfer of responsibility. The
title of the Weapons and Electrical Engineering Branch was then changed to become the Weapons
Engineering Sub Branch of the Engineering Branch.

WE Category Changes

Along with the new ME rates, the rates in the WE Sub Branch were also changed with all
artificers becoming known as Weapons Engineering Artificers (WEA). Apart from the Charge
Chief WEA (CCWEA) title being introduced in place of the Chief WEA, four new categories
of WEA were created with the idea of better reflecting the change in technology boundaries
since the formation of the Electrical Branch and to enable more effective training programmes
to be introduced. These categories were shown as suffices after the title WEA and they were
(Action Data) (AD), (Communications and Electronic Warfare) (CEW), Weapon Data (WD) and
Ordnance Control (OC).

The WEA(AD) was made responsible for surveillance radar and sonar performance, the
processing and distribution of all data by the command management system and the presentation
of that data to the ship's action information organisation. This responsibility included the
performance of the command management system and the associated subsystems such as the
computers, computer peripherals, digital recording facilities and, most importantly, the interfaces
of the system to the rest of the combat system equipment.

The WEA(WD) was primarily responsible for RF and control engineering aspects of guided
weapon systems and gunnery systems. This included both digital and analogue fire control
computers as well as the tracking radars associated with missile control and guidance. Another
important area of responsibility was that of non-radio navigation aids which involved gyros and
both analogue and digital transmission of ship's course, speed and stabilisation data throughout the
ship. While this critical data was being fed digitally into the command systems, separate analogue
transmissions were still being retained in order to ensure the integrity of the information in the
event of computer failure or battle damage when the Command's priority actions moved from
the 'fight' to the 'move' function.[117]

The WEA(CEW) was to focus on radio communications, including satellite communications
links, and the peripheral equipment to those systems, such as cryptographic equipment, teleprinters
and message distribution systems. In addition, the CEW was to take on the electronic warfare
systems which, by the 1970s, were becoming more prolific and complex. ESM equipments, such
as UAA1, were not only detecting and analysing RF transmissions but also classifying platforms

using quickly accessible databases. The ECM equipment had evolved from basic noise jammers into more sophisticated equipment which used electronic techniques to seduce, distract or confuse hostile transmissions, particularly missile target seeking radars. ECCM remained in the ambit of the equipment maintainer, usually a WEA(AD) or (WD). On the domestic side, the CEW was also given charge of the ship's sound reproduction equipment and television systems.

The WEA(OC) was made primarily responsible for gun mountings and turrets, rocket launchers, missile and torpedo launcher systems. Magazines and explosives ordnance were included and it was through the WEA(OC)s, who were invariably to become officers of the quarters, that the WEO, and his designated Explosives Safety Officer (ESO), exercised responsibility for explosives on board ship. The OC role required engineering skills ranging from the hydraulics – still used in equipment such as bomb lifts, winches, ammunition hoists and missile handing areas – to electrical power control used in the 4.5in Mk8 gun.

One thing all WEAs had in common was the need to interface with the command system and ship's services, including power, chilled water and compressed air. While the former was an internal interface with the WEA(AD), the latter was now with the ME department. One of the key areas of EBD implementation was establishing credible interfaces points with the ME Department equipment, and while there were some generally accepted rules – such as a designated fuse in the power distribution system – each WE system was reviewed for the optimum interface points and these were formally documented.

In the WE Sub Branch, mechanic ratings were given new titles of Weapon Engineering Mechanic with suffices of (R), (O) or (C) according to category. Although the (O) numbers were much reduced following the transfers to the ME Department, the retention of three categories provided some continuity of training and they were found to be flexible enough to support the four categories of WEA.

One further change brought in was at the Fleet Chief and CCWEA level where the need for a greater breadth of technical competence was acknowledged by giving these ratings the combined specialist categories of either Action Data Communications (ADC) or Weapon Data Ordnance (WDO). The ADC specialists were drawn from WEAs (AD) and (CEW) with WDOs coming from WEAs (WD) and (OC). On promotion to CCWEA, these men were given appropriate additional training to raise their competence in their non-source specialisation to support their systems level qualification and responsibilities.

September 1982 saw the first apprentices of the new post EBD categories start training in *Collingwood*. In the following year the last *Fisgard*-trained artificer entry arrived at *Collingwood* and, subsequently, HMS *Raleigh* then took over the initial entry artificer training in conjunction with all other new entry training. It was also in 1983 that the mechanician title was abandoned and, subsequently, all apprentices became known as Artificer Candidates.

The Impact of EBD

The impact of EBD and the various studies instigated by the EBWG continued to be felt in the WE world beyond the adjustment of responsibilities and transfer of WE ratings to the ME Sub Branch. Most significantly, the 'Systems Approach' towards weapons engineering was officially adopted and formed the basis for officer and section chief training. In June 1980, *Collingwood* started the Charge Chief Qualifying Course (COQC) having stopped its previous incarnation for chief artificers in 1972. The COQC was designed to give the charge chief a recognised civilian management accreditation and to instil the concepts of systems engineering needed for their new role at sea. This involved training in systems performance analysis and, once qualified, the CCWEAs were sent to ships as the senior rating in charge of the systems within their speciality. Training of the WEO and the Deputy WEO was also adjusted to focus more on system performance on board and they, with one charge chief in each specialisation, were designated and used as systems engineers. This role involved the analysis of system performance and resolution of system level problems as well as advising the Command about systems options in the operational role.

CHAPTER 14

THE ADVENT OF THE MICRO-COMPUTER 1979–1990

This period saw a continued reduction of the Surface Fleet but a remarkable and unexpected increase in worldwide commitments. In 1980 the war between Iraq and Iran required a permanent Royal Navy presence in the Persian Gulf to protect the UK's interests, 1982 saw the Falklands campaign and by 1990 the First Gulf War was imminent.

The Type 22 Frigate

In 1975, HMS *Broadsword* started build as the first of the Type 22 anti-submarine frigate class. The ship was equipped to meet the ASW threat of the fast moving and quiet nuclear submarine in the Cold War environment. The role was fulfilled using the Sonar 2031, a passive, long-range surveillance towed array sonar, and Sonar 2016, a new generation hull-mounted sonar with both active and passive modes of operation. Sonar 2016 employed a number of new features including electronic beam steering and, for the first time, an auto-detection and tracking capability. The number of operators was reduced to two and use of solid state technology and advanced built-in test facilities reduced the maintenance and space requirement and the need for ship services. The detection ranges achieved with these sonars made the ship-launched ASW mortar obsolete and target prosecution was carried out using ship- or helicopter-launched torpedoes. This period saw the introduction of the Sting Ray torpedo, which had a range of around 15,000 yards, compared with the 1000 yards of the Mortar Mk10. The Sting Ray could operate in both active and passive search modes and, once launched, it operated autonomously of the launch platform.

Type 22 fitted with CAAIS Command System and GWS25 Sea Wolf. (Collingwood Photographic Department, © MOD Crown Copyright 2010)

Along with three other Batch 1 Type 22s, *Broadsword* was fitted with the DBA5 variant of the CAAIS command system. This comprised the core Ferranti computer system hardware, similar to previous DBA outfits, and software based on that developed for the Batch 3 Leander Sea Wolf conversions. The radar suite comprised the Type 1006 navigation radar, the Type 968 radar for air and surface surveillance and the Type 967 pulse doppler radar for target indication to forward and aft Type 910 fire control radars used with a dual-headed GWS25 Sea Wolf missile system. Designed as a stand-alone, fully automatic point defence weapon system, the operator manning requirement for Sea Wolf was minimal except for the reload crews needed to replenish the six-barrelled launchers fitted forward and aft. Sea Wolf was considered as an appropriate point defence system given that the ship's primary role was towed array patrol operations. No major calibre gun was deemed necessary for this role and, by now, the gun had lost its credibility against the high speed air threat. The omission of the gun did help to reduce operator numbers and the containerised nature of the Exocet system, fitted to give the ship a surface warfare capability, also contributed towards the lowering of the manpower requirement.

In 1979, HMS *Boxer* started build as the first of six Batch 2 Type 22s that were longer by 50 feet than the Batch 1 ships with a similar combat system fit but enhanced communication facilities. The Computer Assisted Command System 1 (CACS1) replaced the CAAIS system in order to meet a requirement to produce a more comprehensive tactical picture, which included passive sensor data.

In 1983, HMS *Cornwall* was laid down as the first of four Batch 3 Type 22s. Following the disappointing performance of the Type 42 combat system against low-level air attack and anti-ship missiles during the Falklands War in 1982, the Batch 3 combat system was upgraded by the addition of the Goalkeeper Close In Weapons System (CIWS) supported by a 4.5in Mk8 gun mounting and the US-designed GWS60 Harpoon missile system to enhance the surface warfare capability. The Harpoon missile was self-contained in launch canisters mounted amidships on port and starboard facing platforms. Each platform could hold four canisters with the missiles ready to fire and there was no requirement for a launcher crew. Command and control was carried out directly from the operations room and once launched the missile became autonomous following a predetermined flight plan to approach, acquire and destroy the intended target. This increase in the scope of the Batch 3 combat system added to the need for a more capable command system to be fitted, and CACS5 – based on the successful CACS1 software and hardware – was selected.

The earlier Type 22s were fitted with a similar CoGOG propulsion system to the Type 42s and the use of steam propulsion, which had supplanted the sail some 100 years earlier, was fast becoming consigned to history. The later Type 22s were bigger than the Batch 1 ships and in the last five ships of the class the Olympus gas turbines were replaced by Spey gas turbines, a more economical engine which could still meet the requirement for high speed manoeuvres. The propulsion was designated as a 'CoGaG' system as the gearbox arrangements allowed two engines to be engaged on one shaft.

The additions to the combat system turned the Batch 3 Type 22 ships into powerful units but with limited capability in area defence, and this role remained the domain of *Bristol* and the Type 42s. However, from a manpower point of view it was very evident that new technology was driving down crew numbers; *Bristol* required nearly 400 ship staff, the Type 42s around 310 and the Batch 3 Type 22 some 250 for a ship with an equivalent number of systems, if not the same type of capability.

Computer Assisted Command System

In the mid 1970s, Ferranti proposed a replacement command system for CAAIS in the Type 21 which would introduce a distributed intelligent display system connected to a local area network. Sensor information was to be centrally processed, held in a data store and shared across the combat system over a data bus which was also linked to the display network. Rather than upgrade the Type 21, the Royal Navy opted to implement this new generation concept in the Batch 2 Type 22s where the operational requirement demanded the combat system functionality of ADAWs plus the added requirement of processing a passive picture based on large quantities of EW and sonar data.

The system proposed for the Type 22 initially used two FM1600E processors to run the operational program and interface with the distributed processors embedded in the display, weapon and sensor hardware, all placing demands on the shared data store. The advanced functionality made available at the displays included the innovative use of a light pen for menu selection and data input, interactive prompts to support the operator at multi-role consoles and the integration of new generation serial data highways to equipment using software interfaces previously developed for ADAWS. The system operation generated a step change in demand on a data store which was, in real time, having to service the displays and combat system peripherals. This requirement proved to be too much for the two FM1600E processors and the result was a sluggish system performance leading to equipment interfaces being randomly shut down and operator inputs not being immediately processed. The FM1600E processors were eventually replaced by two FM2420 processors operating at twice the speed. This change, coupled with additional display processors to support the man machine interface at the consoles, eventually enabled Fleet operational acceptance of Computer Assisted Command System 1 (CACS1) in the Batch 2 Type 22s in 1992. This milestone was achieved some fourteen years after the requirement was endorsed but at a level of operational capability less than the aspired capability which had crept up during the development programme.

The Batch 3 Type 22 had been fitted with additional combat system hardware and an enhanced command system, CACS5, was developed to meet the increased requirement. The upgraded system was based on the final hardware configuration of CACS1 but with the addition of two

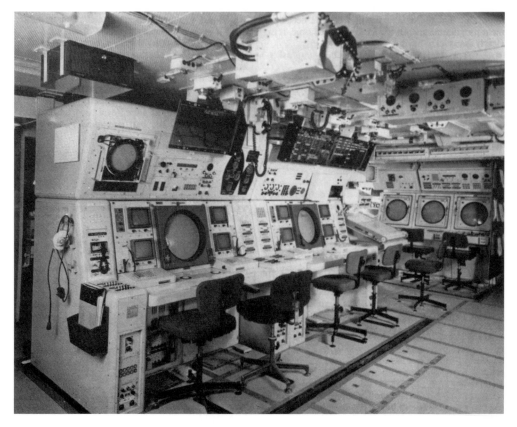

Type 22 CACS operations room. (Collingwood Photographic Department, © MOD Crown Copyright 2010)

more F2420 processors. This architecture introduced the concept of dual processor working, which required more software development but did increase the computer power available and improve the efficiency and redundancy in the system.

Although the birth of CACS just preceded the invention of the microprocessor chip, it did launch the era of distributed processing using embedded mini-computers and this gave a clear sign of the software integration challenges in the years to come. It is significant that budget adjustments early in the CACS programme led to the cancellation of a proposed shore based software development facility for the Type 22. The loss of such a facility meant that the complex software integration task had to be carried out using ship's equipment with all the associated inefficiency and disruption to the Fleet programme. This was a timely lesson for the future as more manufacturers took advantage of the microprocessor to improve the functionality and performance of their equipment. One immediate consequence of the difficulties with the CACS programme was that the contract for the provision of CACS4 into the impending Type 23 Class was cancelled and a new solution was sought.[118]

New Generation Aircraft Carriers

In 1980, the first of three new aircraft carriers, HMS *Invincible*, completed her trials and was accepted into the Fleet, soon to be followed by HMS *Illustrious* in 1982. At one stage in the conception of these platforms they were referred to as 'through deck cruisers' in order to avoid the political issues around the cancellation of the CVA01 in 1966. Initially, the declared operational task for the class was the deployment of ASW helicopters but, with the introduction into service of the naval Harrier aircraft, the façade was dropped and the term aircraft carrier once more became acceptable.

On board these carriers the ADAWS6 command system managed the sensors, including the Type 1022 long range search radar, Type 992 target indication radar, Type 1006 navigation radar, UAA1 electronic warfare suite and Sonar 2016, and controlled the weapons system which comprised the

HMS *Invincible* fitted with ADWAS6 Command System and seen operating an ASW Sea King helicopter. (Naval Museum Portsmouth)

GWS 30 Sea Dart and its fire control radar Type 909. In broad terms, little of the carrier combat system was novel and, as such, the new ships ensured that a lot of manpower-intensive systems would be retained in service for some considerable time until at least the ship's half-life refits.

Type 23 Frigate

The Type 23 anti-submarine frigate was intended to replace the obsolescent capability of the Leanders and the Type 21s. The combat system comprised a comprehensive sonar fit with the new bow-mounted Sonar 2050 for active and passive operation and the 2031 towed array sonar, which was retained until the Sonar 2087 with its active and passive operating modes became available. The ASW weapon carried was the Sting Ray torpedo which was launched either from the MTLS (Magazine Launched Torpedo System) or the Lynx, later the Merlin. Surface weapons included the proven 4.5in Mk8 gun and the GWS60 Harpoon. The primary anti-air warfare system was the GWS26 Vertical Launch Sea Wolf. The missile was launched vertically from a silo thus removing the need for a separate magazine system and to manually reload the launcher. The launch profile also overcame the restrictive blind arc and launcher tracking time problems of the deck-mounted trainable launcher by using vectored thrust techniques to reduce the lateral acceleration forces acting on the missile when it was required to turn hard towards the target bearing immediately on launch.

Another significant feature of the Type 23 platform design was the propulsion system, which was categorised a Combined Diesel Electric and Gas (CoDLaG) system. The electrical drive

Type 23 frigate fitted with DNA Command System, GWS60 Harpoon and GWS26 vertical launch Sea Wolf (Collingwood Photographic Department) (© MOD Crown Copyright 2010)

provided for quiet running during towed array submarine operations with the acceleration and speed needed for tactical engagements made available using the gas turbines.

Together, the new WE and ME technology in the Type 23 was a major contributor to the reduction of the manpower requirement to around 170 crew, down from 250 in a Batch 3 Type 22. There were also commensurate reductions in the shore training costs as, particularly following EBD, the ME Branch started to reap the benefits of having a common electrical training requirement.

DNA(1) Command System

The first Type 23, HMS *Norfolk,* was laid down in 1985 and it was during the build programme that the CACS4 command system contract was cancelled and a replacement was being sought. The DNA(1) command system was selected in 1989 and eventually taken through development by a consortium with BAE Sema as the prime contractor. A distributed architecture was selected based on the use of a combat system highway acting as a data bus to which the majority of the combat system peripherals were connected.

The multi-function display consoles were connected to a local area network and linked to the combat system highway by controlling input/output devices, which also serviced a number of other peripherals including the maintenance facility. The complexity of the software integration problem can be imagined in the light of the fact that some 226 microprocessors were embedded in the DNA(1) system, including the Intel 80386, which was launched in 1985 and a descendant of the ubiquitous 286 microchip that featured in some of the earliest personal computers.

The Falklands Conflict

Three months after Argentinean scrap dealers landed unofficially on South Georgia, on 2 April 1982 Argentine forces began the invasion of the Falkland Islands. Britain's formal response was

HMS *Sheffield* Type 42 destroyer, which was hit by an Exocet missile and sunk in May 1982. (Naval Museum Portsmouth)

RFA *Sir Galahad*, which was bombed and scuttled in June 1982. (Naval Museum Portsmouth)

20mm Gambo close-range cannon. (Collingwood Photographic Department,
© MOD Crown Copyright 2010)

30mm BMARC close-range cannon. (Collingwood Photographic Department, © MOD Crown Copyright 2010)

the sailing of HM Ships *Hermes* and *Invincible* on 5 April to lead a Task Force south to reclaim the islands. This started a campaign which was successful in outcome but costly in terms of ships and men. It also highlighted a number of crucial combat system, ship construction and policy shortcomings in the Fleet at the time. The shortage of close-range weapons capable of engaging missiles and aircraft flying under surveillance radar and missile cover was a particular problem. The scale of the shortcomings was first realised when HMS *Sheffield* was sunk by an air-launched Exocet on 4 May 1982. This was followed by the loss of HMS *Ardent* on 21 May, HMS *Antelope* on 23 May and HMS *Coventry* on 25 May, all due to damage from conventional bombs launched by low flying aircraft. On the same day as the *Coventry*, the RFA *Atlantic Conveyor* was also sunk by aircraft attack. The final loss occurred on 8 June when RFA *Sir Galahad* was so badly damaged by conventional bombs that it had to be scuttled. Apart from combat system weaknesses, experience from the ship losses showed that the lessons of dangers to personnel from splinter, fire and smoke damage inside ships gained during the Second World War had faded and would have to be relearned.

The Admiralty's response was swift in that HMS *Liverpool* and HMS *Southampton* – both at the final stage of build at the start of the Falklands War – were stripped of their newly installed sea boats and given two BMARC twin 30mm gun mountings amidships and two single 20mm Gambo gun mountings on each side of the hangar roof before being dispatched south to join the Task Force.

The experience of the conflict not only resulted in at least the temporary reversal of some of the defence cuts proposed in 1981 under John Nott's Defence Review, but it also initiated a complete reassessment of naval operational thinking with regard to anti-air warfare and damage control procedures.

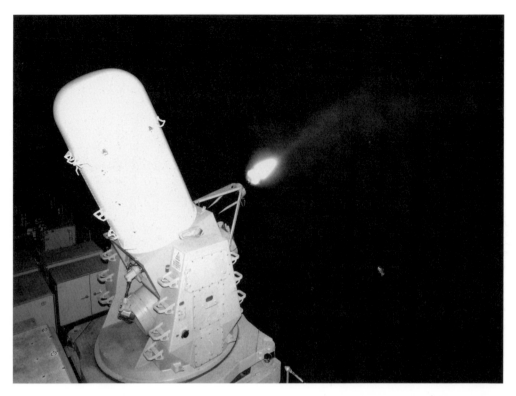

Vulcan Phalanx anti-missile close-in weapons system. (Collingwood Photographic Department, © MOD Crown Copyright 2010)

30mm Goalkeeper anti-missile close-in weapon system fitted to a CVS. (Collingwood Photographic Department, © MOD Crown Copyright 2010)

DNA Command System operator. (BAE Systems Archives)

Weapon Systems off the Shelf

Following the lessons of the Falklands War, a number of new weapon systems appeared rapidly during the 1980s. The operational urgency was such that procurement activity turned towards autonomous, off-the-shelf proven weapons systems with special emphasis on close range aircraft and missile defence.

One of the first examples of such a system was the US Vulcan Phalanx CIWS, which was fitted to *Invincible* and *Illustrious* in the summer of 1982 and, under a rolling programme, the Type 42 destroyers, where it replaced the conventional 30mm BMARC mounting. The Vulcan Phalanx gun was a multi-barrelled 20mm cannon that operated on the rotating barrel principle originally invented in 1862 for the Gatling gun and was capable of extremely high rates of fire. More importantly under the time pressures of hostilities, Phalanx had its own integral target search, acquisition, engagement and damage assessment capability, which meant it required no integration with the ship's command system. As such, the system had the autonomy to attack any aircraft or missile with a flight path determined by the system to be a threat to the ship and operational control was limited to a command veto only. A further ship fitting advantage was that the system only required a 440 volt 3 phase power supply and sea water cooling from the ship's fire main system.

Later in the 1980s, a Dutch alternative to Phalanx, the Goalkeeper, was fitted to two of the carriers and the Batch 3 Type 22s. This system fired a heavier 30mm round at a higher rate and it could simultaneously search and track for targets and carry out continuous threat evaluation such that if a second missile was perceived as a greater threat the system could change targets and prosecute the more dangerous one first. Like the Phalanx, Goalkeeper was a wholly autonomous system, able to search, identify and shoot down targets automatically. They represented a new era in systems operation where the ship input was limited to a command veto rather than an order to engage. Such systems naturally required certain safeguards for friendly aircraft, particularly own ship's helicopter, in the vicinity. Quarters operations requiring human input were limited to reloading the weapons and the quarters emphasis was on the skilled maintainer rather than Gunnery Branch ratings.

The procurement of the US-designed GWS 60 Harpoon surface-to-surface missile system added a powerful surface attack capability when fitted to the Type 23. As with Exocet, the Harpoon was self-contained in launch canisters mounted amidships on port and starboard facing platforms. Each platform could hold four canisters with the missiles ready to fire and there was no need for a launcher crew. Command and control was carried out directly from the operations room and once launched the missile became autonomous following a predetermined flight plan to approach, acquire and destroy the intended target.

The move towards a concept of autonomous and automatic weapons linked to sensor systems only required tactical management at the senior rate warfare supervisor and director level using the DNA command system fitted in the Type 23s. This resulted in further reductions to the number of Operations Branch junior ratings on board ship and more questions about the sustainability of the Branch structure.

Enhancements

While the Falklands experience led to new equipments being permanently fitted into the ships, when the First Gulf War broke out in 1990 a new driver for equipment additions came to the fore and that was interoperability. The need to work more closely with the coalition forces ranged against Iraq meant that the Royal Navy had to acquire equipments that could quickly be fitted to ships moving into the war zone to allow identification and communications between numerous air and sea platforms being used by various national navies operating in the Persian Gulf, particularly the US Navy. With the pressures on equipment procurement budgets, full outfits could not be supplied to all of the Fleet and the concept of 'fit to receive' was

(Collingwood
Magazine Archives)

widely adopted with the key system hardware only being fitted as 'enhancements' during pre-deployment preparation periods.

It was not unusual for equipment to be transferred at sea or appear on the gangway on the day before sailing having been dispatched from the civilian contractor or another ship in another home port, vaguely reminiscent of the wartime RDF and HFDF fitting programmes. This *modus operandi* did keep costs down and did allow some tailoring for the specific war zone duties. Therefore it was not surprising that this approach quickly became a standard operating procedure and, when the Royal Navy was given a role within the NATO forces engaged in the Bosnian Campaign of 1992, the use of enhancement packages continued as the norm.

CHAPTER 15

WARFARE BRANCH DEVELOPMENT

Operations Branch Interface

The Engineering Branch Working Group Study Report, published in 1975, had found that there was a case for looking at the departmental interface between operations and weapons engineering. This came about following what the EBWG had observed to be a decline in standards associated with the custody of weapons ordnance since the introduction of the Operations Branch in 1975 and the increasing commitment to weapons operation and performance assessment by the WE Department. The report had recommended a further study in this area and debate continued well after EBD implementation but with little progress, and at one stage it looked as if the matter would not be resolved although study work was still ongoing.

In 1988, new arguments for change started to appear and it was postulated that the seaman specialist did not need the tactile and cognitive skills of his predecessors because of advances in technology. These advances, it was suggested, would make the maintenance and repair task much simpler and this presented an opportunity to provide some maintainer training for the Operations Branch junior rate, which would broaden the base of the operator's job. Not only would this make the job more attractive to new recruits but it would also allow a reduction in the number of dedicated maintainers required on board ship.

Others countered this argument with the fact that new technology had clearly demonstrated that the number of operators at the junior rating level had been reducing since the 1960s owiing to the increase in automated data processing. This was going to continue for the foreseeable future when eventually the junior rating operator was not going to be needed at all and all command information would be presented through the command system software in terms of veto or priority options. Thus the real issue was the fact that technology was making the Operations Branch structure unsustainable in its current format, i.e. no junior ratings as a source for the senior ratings that would still be required in the operations room of the future. Whether the introduction of a user maintainer was actually a viable way of managing Operations Branch sustainability was felt to be open to question, given the wide variation of technological advances in the Fleet. Another factor against the proposal in the minds of many WEOs was the very real possibility that any change that altered the Ops/WE balance would lead to the WE Sub Branch structure also becoming unsustainable.

Other Changes

For whatever reason, technological pressures were creating a problem for rating recruitment to the Operations Branch as the scope of careers was limited and perceived as becoming less fulfilling in the long term. After the fall of the Iron Curtain, the calls for a peace dividend and the resulting publicity given to redundancy programmes in the Armed Forces also started to take its toll on recruiting across all arms of the Service. Although not publicly conceded, it was felt by many that it was the recruiting problem that led to the next significant change in the Royal Navy as a whole, not just at the Ops/WE interface.

(Collingwood Magazine
Archives)

That momentous change came on 5 February 1990, when a decision was announced in the House of Commons to extend the employment of members of the Women's Royal Naval Service (WRNS) to include service at sea in surface ships of the Royal Navy. Accordingly, their conditions of service were altered to allow volunteers to go to sea. The first operators were sent afloat on 8 October 1990 and the first ship to have mixed manning was the Batch 1 Type 22 frigate, HMS *Brilliant*. In September 1990, the first female artificer apprentices and mechanics were recruited and entered HMS *Raleigh*. By 1994, all major surface classes of warship had women serving on board and only the minor war vessel and submarine flotillas retained all male crews. In 1994, the Women's Royal Naval Service was abolished as a separate entity within the Royal Navy and all female officers and ratings discarded the traditional blue stripes and badges and, where appropriate, took up the gold lace and gold badges.

The year 1993 saw yet another change that was to be highly significant and the forerunner of events over the next ten years which would alter the nature of *Collingwood* as the Weapon Engineering School of the Royal Navy. This change was the transfer of training for the Communications Sub Branch of the Operations Branch from *Mercury*, its home since the Second World War, into *Collingwood*. This came about as part of the rationalisation of the Royal Navy's training estate which was being pursued vigorously as part of the cost savings arising from the reduction in manpower numbers following the end of the Cold War.

Almost unnoticed during this period, the title of Fleet Chief Petty Officer was abolished and replaced by the rank of Warrant Officer. This ended the debate that had been going on since the inception of the Fleet Chief Petty Officer as to why, on tri-Service grounds, the term Warrant Officer had not been used in the first place, particularly as the Fleet Chief had been entitled to the salutation of 'Sir' from his subordinates.

Warfare Branch Study Recommendations

In 1992, after much debate, the final report arising from the EBWG recommendations, the Ops/WE Interface Study, was published. The report recommended that the existing Operations Branch should be re-branded as a new Warfare Branch. The main reasons given for the reorganisation were increased automation in data processing and improvements in weapon equipment reliability which had reduced the required number of sea billets for Operations Branch junior ratings and WE Sub Branch weapon engineering mechanics, both senior and junior ratings. The assurance given by the supporters of the change was that Warfare Branch structure would be designed to ensure better career opportunities and prospects for its members than those available under existing Operations and WE Sub Branch arrangements. Also, more importantly, the new structure would better meet the operational needs of the Fleet and, with that in mind, Navy Board approval was given to implement the change.

The new Warfare Branch was to be formed by transferring all WE mechanic ratings over from the WE Department to a new Warfare Department where they would become operator/mechanics and be responsible to the warfare departmental officers, many of whom had been principal warfare officers in the Operations Branch. Operations Branch ratings were also given the opportunity to volunteer for transfer to the Warfare Branch and take on a maintenance role. Recruiting for the WE Mechanic stream and the Operations Branch ceased in 1993.

WBD Implementation

The result of WBD implementation was a reduction of the WE Department, which left the WEO with the management of only the artificers in the complement, many of whom were weapon and sensor operators in a quarters position at action or defence stations and under the control of warfare officers in those roles. When equipment required deeper levels of skilled maintenance or repair, the artificer was called in to work with the operator/mechanic who reverted to the artificer's technical control at this stage in the task.

In the spring of 1993, the first ratings for the Warfare Branch were recruited. The ratings, both male and female, were given the title of Operator Mechanic (OM) with a number of sub-specialisations open to them. Depending on specialisation, embedded in the OM's tasking was the requirement to operate weapon systems and support the WE artificer in maintaining such systems in much the same way as their WEM predecessors. In many ways this was a return to the same working structure which pertained up until the end of the Second World War, when the LTO, who had some electrical training, worked for the Torpedo Officer. The only difference was that in 1944 the Torpedo Officer also employed the electrical artificer and, arguably, this provided a more coherent structure.

The Operator Maintainer

The new entry OM was given eight weeks basic training at *Raleigh* before commencing seven to nineteen weeks professional training at various establishments depending on category. On completion of specialist training they were sent to sea as an OM 2nd Class (OM2), able to operate their specialist equipment and carry out minor maintenance under the supervision of the section WEA. Their advancement path was to OM 1st Class (OM1) and then on to Leading Operator Mechanic (LOM). Advancement was geared to learning more about the technical support of their equipment and its operation. The key objective was to bring them up to a level where they could carry out semi-skilled maintenance and repair activities which were comparable to that of the old WEM and LWEM but the range of their equipment knowledge was more limited. The management of maintenance and repair of equipment remained as the responsibility of the section WEA. Warfare Branch ratings could also be attached full time to section WEAs to gain experience in the specialist care and repair of equipment. Once the rate of petty officer and above had been achieved, OM employment was more focused on their warfare duties in support of the PWO and the day to day management and supervision of OMs in their parent warfare department.

The new branch structure posed a number of problems with promotion prospects, one of them being that the OM did not have a viable career path through to commissioned rank as a WE officer. While selection for WE artificer, and thence to the WE SD List, was theoretically open to them, the rate at which technical breadth of experience could be gathered by the OM was significantly slower than their WEM predecessors. This made it difficult for them to achieve the standard required for artificer candidate selection. On the other hand, the WEMs and LWEMs who transferred were better placed to take advantage of the opportunity for artificer but everything was dependent on the scope of their new job inside the warfare department. Matters were not helped by the fact that their divisional officer was no longer a weapons engineer and, therefore, unaccustomed to identifying artificer potential, a situation which was unlikely to improve.

(Collingwood Magazine Archives)

Operations Branch junior ratings who joined the Warfare Branch and wished to further their careers as PWO specialists could also achieve this via the SD List, but the more junior the rating the less likely they were to gain sufficient warfare experience for selection. Apart from the problem of getting this experience, the competition for promotion was made much more intense by the transfer of the WEMs into the new branch. Many of these WE ratings were better placed by way of their educational background, naval technical training and experience to take advantage of any promotion opportunities to senior rate and officer in the Warfare Branch. Consequently, the career prospects for transferred Operations Branch ratings below leading rate were adversely affected and the potential feed into the artificer candidate training scheme was also reduced.

Warfare Branch Categories

The Warfare Branch specialist operator maintainer categories on board major warships were aligned with various areas of warfare specialisation namely Above Water Warfare, Under Water Warfare and Electronic Warfare.

The Above Water Warfare ratings, OMs(AW), were responsible for compiling the air and long range surface picture in the operations room using data received from onboard sensors and other units. They also manned the ship's guns and missile systems and looked after the associated stores and small arms. At the LOM level, these ratings specialised further into the above water tactical (AWT) or warfare (AWW) fields. The sources for the interim manning of the AW specialisation branch were the radar and missile categories in the old Operations Branch or the ordnance and radio categories from the WE Branch.

The Under Water Warfare ratings, OMs(UW), were responsible for the operation of passive and active sonar systems and maintaining a sub-surface picture plot. They were involved in submarine and torpedo contact procedures and acted as surface picture compilers with a knowledge of tactical data and radar systems. The ratings considered suitable for this new rate were WEMs(R) or (O) and ABs (Radar) or (Sonar).

The other OM categories were in the more specialist fields of electronic warfare (OM(EW)), communications (OM(C)) and mine warfare (OM(MW)). While the duties of the Operations Branch personnel who worked in these areas changed little, WEMs who were transferred found that the scope of the work that they now had was greatly reduced in both technical and operational content.

As far as the WEO was concerned, the loss of the WEM structure was significant in that the day to day technical management rested in another department and only when the problem reached a critical stage was it referred to his notice. As quarters officers, many of the WEAs were more involved with the warfare department than their own. To make matters worse, accountability for technical standards was not linked to control of the resources needed to maintain those standards and this represented a fundamental management system weakness that was not welcomed by most in the WE Sub Branch.

Further Warfare Branch Development

In July 2003, the concern for the sustainability of the WE rating structure was sufficient for the Navy Board to call for a study, the Fleet Competences Work Strand Study, which concluded that the WE and Warfare Branch interface was not working as well as had been hoped. The WEA structure was proving to be unsustainable partly because of recruitment shortfalls and moratoria but also because the anticipated stream of OMs coming into the WE Sub Branch was not forthcoming. Reasons for this included the fact that many of the new entry OMs were unable to meet the academic standards required for artificer and, even if they were eligible for selection, they were still limited in their technical training and experience. Another factor was that those OMs who were academically suitable for posts as artificers, which included many of the transferred WEMs, felt that their career prospects were better if they remained in the Warfare Branch, potentially as officers. The latter situation provides an interesting parallel with the weakness of the Selborne-Fisher officer training scheme in the early twentieth century where, at the end of a lengthy common training period, when asked to volunteer for a specialist branch, few elected for the Engineer Branch because they could not achieve sea command which was still a prerequisite for many senior posts in the Service.

Other reasons published in the Royal Navy's Broadsheet[119] for 2004 were that OMs were doing too much multi-tasking and that WEA artificers was not getting the continuity of technical support to complete routine maintenance. In short, the WEA was dependent on the Warfare Branch for support with day to day tasks and, because this was not always forthcoming, the lack of a dedicated WE section junior rating was affecting the performance of the WE Department. The parallels with the pre-WBD employment of WE seamen were now made obvious but this experience had been ignored, along with several other significant lessons learnt during the development of the technical branches over the previous 90 years.

In May 2004, the Navy Board approved the recommendations of the Fleet Competences Work Strand Study in principle and directed that the Warfare and Engineering Branch Development Management Groups should complete the preparation work for restructuring both branches and to take it through to implementation with the following remit.

The new Warfare Branch structure was to comprise three specialist categories, namely warfare, communications and information, and seaman. The replacement structure was to be more focused on warfare tasks involving weapon, sensor and information management to meet the new operating requirements of the Astute Class submarine, the Type 45 destroyer and the planned new carriers, which were in the early stages of procurement. Seamanship tasks were to be separated from the warfare roles and given to the newly created seaman specialist.

Initial proposals were that warfare specialists should be given further sub-specialist qualifications in radar (AWT), sonar (UW), gunnery (AWW) and electronic warfare(EW). Communications and information specialists were to continue with their communications functions and take on the management of information systems being used in the Fleet. In conjunction with this change, all equipment maintenance aspects were to become the responsibility of the WE Sub Branch with the transfer of maintenance tasks being dependent on concurrent and commensurate changes being made under the auspices of the Branch Development Management Group for the Engineering Branch.

Further Engineering Branch Development

Broadsheet 2004 also announced the changes to the Engineering Branch, which were to be carried out concurrently with the restructuring of the Warfare Branch. The announcement declared 'rapid advances in ship system technology and the merging of weapons and marine engineering technologies, combined with the need to tackle the shortage of artificers, have led to a fundamental review of the Engineering Branch.' This review had recognised that a two-tier artificer and mechanic structure was no longer suited to the support of new technology and it recommended that all ratings in the Engineering Branch, both WE and ME, should have a single career path within their sub branch.

For the ME Sub Branch this meant the merging of the MEA and the MEM structures to form an Engineering Technician (ME) (ET(ME)) structure which offered a progressive career path for all new entry recruits to commissioned rank. In the case of the WE Sub Branch, the mechanic base needed for a similar ET(WE) structure was to be formed, initially, from Warfare Branch OMs being transferred into the WE Sub Branch simultaneously with the total equipment maintenance task responsibility being taken on by the WEO.

The first new entry ET(WE)s entered *Raleigh* in January 2006 and joined the Fleet in August 2006 after completing a 17-week ET2 qualifying course at *Collingwood*. On completion of task book training, the ET2 became eligible for advancement to ET1. In due course, career qualifying courses to Leading ET (LET) and Petty Officer ET (POET) would bring the standard of training up to a level previously achieved by a POWEA on confirmation in a first sea job in charge of a maintenance section. Most importantly, completion of the POET qualifying course was accredited as a foundation degree which, within a lifelong learning context, had become arguably more recognisable to industry than the artificer apprenticeship.[120]

EPILOGUE

When I was researching *The Greenie* and trying to define parameters for the book, I asked numerous individuals what they would define as a Greenie. Was it simply a member of the original Electrical Branch? Post Engineering Branch development, did it apply to the Marine Engineering Sub Branch as all marine engineering artificers and mechanics received electrical training? With the introduction of the Warfare Branch, the operator mechanic also received an element of electrical training. Were they also Greenies? Looking back, the same question could be asked about the seaman torpedoman and the torpedo artificer; but it would seem unnecessary to ask the question about the electricians, electrical artificers and the wartime electrical mechanics whose credentials made them clearly 'Greenies' but who worked for the Torpedo Branch.

One response to the question was 'anyone who has held a screwdriver for the purpose of maintaining a piece of electrical equipment.' This seemed a very inclusive definition and, given the events of EBD and WBD, I am not certain that the *esprit de corps* associated with the term 'Greenie' was used with the same affection by the hived-off elements of the Electrical Branch in its subsequent guises.

The most recent restructuring, with suitably trained personnel specialising in WE, ME and AE technology areas, looks very like the vision of the Engineer-in-Chief of the Navy in 1944. It reinforces the single Engineering Branch concept and channels the contribution that engineering can make to the future of the Royal Navy more effectively than with the three previous sub branches, with one seriously weakened by the loss of the semi-skilled mechanic ratings under WBD. The new Engineering Branch should provide a streamlined engineering technician base which can overcome much of the discontinuity and redundancy in training that was present in the previous structure, driven largely by the existence of the artificer and mechanic rating categories. In the new branch, all ETs will inevitably receive electrical training; are they, therefore, all now to become qualified as Greenies? Of course there is no real need for an answer, and hence the decision to base the book around the technology, rather than a specific group or period in time, and give anyone who can relate to the technology an opportunity call themselves a Greenie 'if the cap fits'.

With Admiral Watson's assessment of the early days of the Electrical Branch in mind, it seems appropriate to carry out a similar appraisal of events since 1946. As this will be a personal view, taken by someone with a less exalted background than Admiral Watson, before giving such an appraisal it may be best to declare the author's perspective. Having joined the Weapon and Radio Engineering Branch after graduation, I switched to the Weapons and Electrical Engineering Branch before appointments as Deputy WEEO in HMS *Ajax* followed by WEEO in HMS *Falmouth*. In both posts I gained considerable experience of power generation and distribution systems. I was subsequently converted to the Weapon Engineering Branch and found myself on C-in-C Fleet's Engineering Staff as the representative WEO on the Fleet EBD implementation team. As such, I was one of a team of three which took the EBD word to every ship in the Fleet and controlled the transfer of high power sections etc to the ME departments.

After Fleet Staff, I joined HMS *Newcastle* as WEO in 1983. The ship had been EBDd and this is where I gained some first-hand experience of the way things were working out at sea. On promotion, somewhat perversely, I was appointed as the Naval Weapon and Electrical Overseer on the Tyne with a civilian staff who were clearly not EBDd and in fact they had only just combined together to form a weapons and electrical department. Thus I found myself again responsible for power generation and distribution system installations in new build ships. I was then appointed to *Collingwood* as Application School Commander, where I became closely involved in the debate over the WE/Ops interface. With the Warfare Branch imminent, my last job in the Royal Navy was as Squadron Weapon Engineer Officer to the 5th Destroyer Squadron, where I was able to gauge the reaction of many WEs to the proposed structure changes under WBD. On leaving the Royal Navy in 1992, I joined Ferranti Simulation and Training working in the field of command team training simulators. As my job heavily involved warfare training, this put me in an ideal position from which to observe subsequent Branch development activities and, more closely, the progress of technology in the naval warfare field.

The timing of EBD could not have been much better; the gas turbine propelled ships were quickly coming into service with most of the Type 21s, *Bristol* and two or three Type 42s undertaking the departmental reorganisation while operational. The bulk of the Type 42s and the Type 22s were therefore organised on EBD lines during the build stages. A great deal of preparatory work was done to redefine technical responsibilities and interface boundaries between the weapons and marine engineering departments. The eventual WE interfaces were much more simple than under the previous regime because the maintainers, documentation and the general working environment were suddenly grouped to support a complete system which, if any, had logical interfaces to the ME, or occasionally AE, systems in the ship. In the WE case, the most predominant interface point was the power supply fuses for the equipment.

More importantly, the personnel involved appeared to be very comfortable with the proposed changes, notably all the WE officers who transferred to the ME Sub Branch were volunteers and the number of pressed ratings was very few. Generally, I believe this was because their job specification had not materially altered. As far as the ship's complement was concerned, the High Power Team was moving as a section of artificers and mechanics and liaising with the rest of the ME department only slightly more closely than they had been previously. As the electrical experts, they also had access rights to the MEO through a chief artificer from the WE Branch who had transferred to become a Chief Marine Engineer Artificer with an electrical specialist suffix of (L), thus CMEA(L). The CMEA(L) also held parity with the Chief Marine Engineer Artificer who was recognised as being non-electrically trained by having the suffix (P) in his title: CMEA(P).

The main concern for most WE transferees involved ME watchkeeping duties and the need to gain certification before being eligible for advancement. Few senior rate transferees wanted to undertake what was a normal way of life for the senior rate marine engineers. This issue was dealt with by placing no obligation on the transferees to undertake marine engineering training for watchkeeping in order to achieve advancement. After transfer, they were only required to keep watches in the switchboards and conduct electrical equipment rounds where they had previously done so as WE senior or junior rates. As there were few changes to the total numbers in the ship's scheme of complement, no one from either the (P) or the (L) side was any worse off in terms of watchkeeping after the transfer. My experience as WEO of *Newcastle* convinced me that the reorganisation had been successful and by then the electro-mechanically trained MEAs and MEMs were joining the ME department.

The WE/operations department interface presented a very different problem and this was reflected in the time taken to reach conclusions and implement any findings. In fact, at one stage in the mid 1980s the study work seemed to be taking a low profile and the idea seemed to have been put on the back burner. The issue that kept the topic alive was that advancing technology was continuing to reduce the number of junior rate positions in the operations department and this was fuelling concern that the manning structure of the department was going to become top heavy and unsustainable.

An argument was put forward that technology would provide an answer in that every operator would be able to carry out maintenance duties because, as a generation, the recruiting base would be more computer literate. It was suggested that these indigenous skills, along with suitably

(Collingwood Magazine Archives)

designed systems, would allow the implementation of a user/maintainer trained rating at little extra cost. As there was already a shortage of junior ratings in the Operations Branch, it was also proposed that existing WE mechanics, both senior and junior ratings, would be a ready-made group who could fill the short term requirement for operators and that, once fully associated with Operations Branch matters, they would become ideal candidates to fill the director and supervisor posts in the operations room.

This proposition underestimated two fundamental aspects, the first being the nature of the WE mechanic, and secondly the real rate of progress in technology implementation in the Fleet under the post-Cold War priorities being imposed. A further issue was the impact on the sustainability of the WE artificer structure.

The importance of the nature of the mechanic and its relationship to the proposed changes was discussed but understated at the time, possibly because of the semi-skilled status accorded to the category. In actuality, although the WE mechanic was classified as being semi-skilled this was belied by the quality and motivation of the average recruit. The NAMET requirement was higher for WEM than for the Operations Branch entrant and this inevitably introduced an element of selection in the recruiting office. Also the majority of recruits had made the decision to enter as a WEM on electrical engineering interest grounds. This is borne out by the significant number of mechanics wanting to come through *Collingwood* as artificer candidates. It was certainly the case in the 1980s during my own tenure as Commander Application Training School, when the passing grades achieved by artificer candidates were more than comparable with their artificer apprentice colleagues.

An aggravating personnel issue was the resulting position of the WE officers and senior rate artificers having lost their WE mechanics to the Operations Branch. Without any direct control over and a reduced mentoring relationship with both senior and junior rate mechanics, there was a WE Sub Branch perception that this would affect not only the mechanic but also the fulfilment that many artificers felt as a result of nurturing the young mechanic in the learning of his trade. Unfortunately, this perception was not considered significant by the proponents of the change.

Another factor, not given enough weight in the debate, was that the precedent for the type of working relationship being proposed had been tested previously. This precedent will be well remembered by ship WEOs of the time and it was the employment of seamen in the WE department as designated WE seamen. This employment was undertaken on a rotational basis by operations department seamen with three or six months being the period of loan. When the period was too short it was highly disruptive to lose a man who had been trained up as a maintainer. When the time on loan was extended, a significant number of WE seamen put in requests to transfer to the WE Branch. Although these ratings were attached to a WE section, usually underwater weapons or surface weapons, they were still liable for seaman duties. Those duties had priority over WE activities and did lead to disruption in the running of a section. In a similar vein, many WE senior rate artificers had worked for the operations department as quarters officers. As such, they had experience of the conflicting demands that this situation could place on them and they could see the problem if they lost line management control of their supporting mechanics.

Other inconvenient questions involved the technology and how much it would make the operator maintainer, primarily at the junior rate level, a viable proposition. There had been signs that the number of required junior rate operators had been falling since the introduction of the digital computer in the 1960s, a classic example being in the sonar field where *Ajax* had required eight junior rate operators to man the two main sonars, Type 170 and Type 177, in action. This manning requirement was reduced considerably with the introduction of Sonar 184 and then Sonar 2016. Eventually the Sonar 2050 came into service and required only a single WEA operator maintainer, whose primary role was to oversee the functioning of the system not to participate in the tactical operations. The question was, therefore, why was technology only now a reason for branch restructuring? Looking back, there was a simplistic view that with the younger generation avidly playing Pac Man and Space Invaders and having access to the 286 microchip PC, these gaming and keyboard skills plus user friendly 'Help' facilities would make every recruit a computer whiz kid on joining. In fact, 'Help' facilities would make fundamental technical knowledge virtually redundant allowing maintainer functions to be carried out with the minimum amount of training.

Notwithstanding the potential for technology assisting the support function, it was still a big leap of faith at the time as, even then, advances were dependent on the naval ship programme proceeding to schedule with the same amount of political commitment as during the preceding Cold War years. With the benefit of hindsight, this did not happen, with the Type 23 failing to make the anticipated progress with its command system and, consequently, prolonging the need for ad hoc operation of the combat system almost until the end of the 1990s. Up to that point, the only real contribution to operator numbers reduction made by Type 23 Class new weapons technology was the Sea Wolf vertical launch system and the Magazine Torpedo Launch System both of which reduced the need for magazine maintenance, operations and upper deck loading crews. Apart from this, the Type 23 peripheral equipments to the command system were little different to that in the Batch 3 Type 22s and demanded the same number of maintainers and levels of skill.

Also in the early 1990s, the Royal Navy started seeking a replacement for the Type 42 and options for cooperative procurement were being investigated, reinvestigated and eventually rejected until the compromise leading to a UK Type 45 was reached. This meant that some 20 years had passed since the Navy had last taken delivery of a new major warship class. In that time electrical technology had leapt forward with cheaper COTS technology proliferating, but had it reached the point where the Navy could realise the training and manpower savings which had been envisaged as achievable under the Warfare Branch user/maintainer concept? Evidence from the little success many people experience with embedded 'Help' facilities should, perhaps, have made the Royal Navy more sceptical about such a facility being the panacea for fewer men with less training. Certainly experience in the simulation industry tells me that for front line operations such support facilities have many shortcomings. Without dwelling further on the reasons why it came about, in April 2007 the Engineering Technician structure was brought into being.

At the risk of being presumptuous and having only limited information at this time of writing, it would seem that the new Engineering Branch structure looks pragmatic and addresses many of the concerns that were debated, but dismissed, in 1993. The new ET structure seems to go further and offer a coherent technical development programme, which should be far less

(Collingwood Magazine Archives)

disruptive to ship manning and training processes to the Royal Navy's benefit and to career progression. The achievement of a foundation degree underpinned by years of relevant experience and training should also shine as beacon amongst the plethora of competing civilian awards, which seem to be more cognisant of the education system than the quality of the outcome. This aspect, at least, should be able to stand comparison with the respect accorded to the WE artificer training of the past.

One final point worth mentioning is that most ship technology now and for the future will involve electrical technology. If this is the case, then the new Engineering Branch can now justifiably lay claim to being the new 'Greenie' branch and keep the ethos going.

GREENIE PEOPLE

The Second World War brought together a broad spectrum of people with a common interest, aptitude and skill for working in the specialist fields that were rapidly evolving under the umbrella of electrical engineering in the Royal Navy. Many were regular service personnel with electrical experience gained under a disparate pre-war organisation which had not fully grasped the demands of the emerging electrical technologies. Others had initially chosen their careers outside the Royal Navy and had been trained in a civilian environment before fate led them to serve their country in their chosen profession. The timing of the formation of the Electrical Branch was apposite in that many people who would otherwise have lived out their lives outside of the Service now had a glimpse of working at what was the cutting edge of this exciting new technology. It was this experience and the opportunities offered in the new branch which persuaded many to stay on in the Royal Navy after hostilities ended. This chapter is intended to illustrate the huge effort made to meet the ever-changing technical requirements for fighting a world war and to celebrate the diversity, determination and sense of duty shown by the people who participated in the Royal Navy's contribution to such a formidable venture. Some contributions offer a detailed view of the organisation as it rapidly evolved from the period of the Phillips Report onward, and others present a quick snapshot of their service lives during those turbulent times.

Rear Admiral S.L. Bateson CB, CBE

Rear Admiral Bateson joined the Royal Navy in 1916 and carried out his cadet training at Keyham College followed by a midshipman year in HMS *Marlborough*. He was then appointed to HMS *Walpole* in which he went to the Gulf of Finland in support of a raid on Kronstadt Harbour in 1919. While on the experimental staff of HMS *Vernon,* he was involved with mine development trials. This work included the conversion of HM Submarine *M3* into a minelayer and the alteration of the stern of the cruiser minelayer HMS *Adventurer* to make her capable of laying mines at high speed. After serving at the Admiralty in the Directorate of Torpedoes and Mining, he was appointed Fleet Torpedo Officer Mediterranean on the staff of Admiral Sir Dudley Pound.

As the Commanding Officer of HMS *Latona* he undertook a record-breaking journey from the Clyde to Alexandria via the Cape of Good Hope, completing the 15,258-mile journey in 31 days. From Alexandria, *Latona* made twelve successful runs to Tobruk in support of the Army, at times being obliged to unload

Rear Admiral (L) S.L. Bateson CB CBE, the First Electrical Flag Officer. (Collingwood Museum)

some 150 tons of cargo in 55 minutes under air raid conditions and at night. On 25 October 1941, while on her thirteenth trip, she was sunk by what was believed to be the only successful night attack by a mixed force of German and Italian bombers.

After an appointment as Flag Captain to Rear Admiral Rawlings in HMS *Ajax* which, as a light cruiser, at one stage was the heaviest unit in the Mediterranean owing to the severe losses sustained in that area, he became Naval Assistant to the Controller of the Navy involved in provisioning for the Pacific War. On 3 September 1945, as the Commanding Officer of HMS *London*, he received the surrender onboard of the Japanese Forces at Sabang one day before the general surrender of Japan.

On return to the UK, Rear Admiral Bateson was appointed as a Captain to be the Director of the Naval Electrical Department and given the task of establishing the Electrical Branch in accordance with the Middleton Report recommendations. In 1949, Rear Admiral Bateson was given recognition of his achievements for the Electrical Branch when a new naval rank was created and he was promoted to Rear Admiral(L) and became the first Electrical Flag Officer.

Vice Admiral Sir Philip Watson, KBE, LVO

Vice Admiral Sir Philip Watson KBE LVO. (Collingwood Photographic Department)

Admiral Watson was initially trained as an electrical engineering student apprentice with the London, Midland and Scottish Railway Company. He joined the RNVR early in 1940 as an Electrical Sub Lieutenant, first went to sea in HMS *Seagull* in August of that year and subsequently served in HMS *Hebe* (First Minesweeping Flotilla), HMS *Nelson* (as Assistant Torpedo Officer) and HMS *Berwick* (as Torpedo Officer). In January 1946 he transferred to the Royal Navy as a Lieutenant and his first appointment was as Naval Assistant to Rear Admiral Bateson who at the time was charged with forming the Electrical Branch.

After completing the long radio course at *Collingwood*, he joined the staff of the Captain 5th Destroyer Flotilla, serving in HMS *Solebay* and HMS *Gabbard*. This was followed by a period at the Admiralty in London before returning to *Collingwood* as the assistant to the training commander. As a Lieutenant Commander, he then served in the radio section of Malta Dockyard and as Electrical Officer in HMS *Decoy* before being promoted to Commander (L) in 1955. After an appointment in the Admiralty, he became the Electrical Officer of HM Yacht *Britannia* after which he was made a member of the Royal Victorian Order. In 1959, he moved to Chatham dockyard for three years in charge of the electrical shops and the weapon section before again returning to sea in 1962, this time as the Weapon Electrical Engineer Officer of HMS *Lion*.

Promotion to Captain took him to the Ship Department at Bath where he was involved with submarine, aircraft carrier and commando ship design. From 1967–1969 he was the Captain of *Collingwood* before returning to Bath as the Deputy Director of Engineering (Electrical) in the Ship Department. He was promoted to Rear Admiral and became the Director-General Weapons (Naval) before becoming the first Electrical Branch officer to achieve promotion to Vice Admiral. Admiral Watson was awarded the KBE in 1974 for his distinguished service before retiring from the Royal Navy in March 1977.

Commander H. Boyce DSC RN

Commander Boyce joined the Royal Navy in November 1940 as an Electrical Sub Lieutenant RNVR and after an indoctrination course at *Vernon* he was sent to the minesweeper base in North Shields.

In 1941, he was one of seven sub lieutenants sent to various Commonwealth navies to supervise the installation of minesweeping equipment and to provide instruction in magnetic and acoustic minesweeping equipment. Commander Boyce was sent to Simonstown in 1941 where, as well as the fitting out of the ships and instructional duties, he became responsible for rendering safe mines which had been washed ashore in the Cape area. In 1943, he returned to join the staff of Captain Hopper who was in command of the minesweepers in Queenborough.

Shortly after D-Day, General Montgomery could not get sufficient supplies to support the Army's rapid advance, therefore he ordered Admiral Ramsay to clear the River Scheldt of enemy mines and open up the route to Antwerp within three weeks. Lieutenant Commander Boyce was dispatched to ascertain the damage to the port of Vlissingen at the entrance to the Scheldt and in one of the pill boxes discovered a chart of the German minefields in the river mouth. This resulted in a large minesweeping force being deployed and over 400 mines were destroyed but sadly with the loss of several minesweepers. Antwerp was opened up within two days of Montgomery's deadline, although in the minesweeping force fingers were crossed whilst the first convoy proceeded up the river. Captain Hopper was awarded the DSO for the operation and Commander Boyce was awarded the DSC.

The day after VE day, Commander Boyce was in HMS *Prompt* sweeping mines off the Hook of Holland when the ship was blown up by an acoustic mine. He survived to be offered a permanent commission in the Royal Navy as an Electrical Lieutenant Commander and after a short course he was appointed to HMS *Belfast*, the flagship of the British Pacific Fleet in Hong Kong. He was the first electrical officer to relieve a torpedo officer of his electrical duties and was designated the Pacific Fleet Electrical Officer. His next job was First Lieutenant in *Collingwood* before promotion to Commander and an appointment to HMS *Nigeria* in the South Atlantic Squadron. A year later he transferred to HMS *Bermuda* when she relieved *Nigeria* in Simonstown. After a spell with the Reserve Fleet in Portsmouth, Boyce returned to sea onboard HMS *Theseus* as the Squadron Electrical Officer. Finally with sea service behind him he went to Chatham dockyard in 1951 to be involved with cruiser and destroyer refitting. This was followed by a period with the nuclear division in Bath engaged in preparing the Royal Dockyards for nuclear refits. Commander Boyce retired in July 1966.

Commander C.A. Cambrook RN

Commander Cambrook joined the Service in 1944 as a Probationary Temporary Acting Midshipman (Special Branch) RNVR borne for radar duties. After qualifying as a radar officer, he took passage in RMS *Mauritania* from Liverpool to Sydney Australia, westwards across the Atlantic and Pacific Oceans, and joined HMS *Grenville* as the Flotilla Radar Officer to the 8 'U' Class destroyers of the 25th Destroyer Squadron.

In 1946 Sub Lieutenant Cambrook served in the cruiser, HMS *Euryalus*, in Hong Kong

Defiance Electrical Training School. (Plymouth Naval Museum)

before returning to England in 1947 and joining long electrical course 33L7 at *Collingwood* and becoming one of the founder members of the Electrical Branch. At the time, some of the long course training was carried out in *Defiance,* which then comprised three hulks moored in the Hamoaze opposite Devonport dockyard.

Lieutenant Commander H.E. Coleman RN

Lieutenant Commander Coleman returned to the UK onboard HMS *Woolwich*, a destroyer depot ship, after a period in the destroyer workshops in Trincomalee, Ceylon. He joined *Collingwood* as a regulating chief petty officer and soon heard that there were openings for EAs with suitable aptitude to take a cross-training course for radio electrical artificer. He applied for the course and, despite being at the advanced age of 30, he was accepted. On course he showed himself to be capable of absorbing the technology and passed out top of the class to become the first Chief Radio Electrical Artificer in the Royal Navy.

CREA Coleman successfully took the promotion examination for Commissioned Electrical Officer(R) and was promoted in 1948. He was then sent to HMS *Triumph* where he served for two years in the Mediterranean. This was followed by a period in the Far East that included the Korean War.

Lieutenant David Deacon RN

Lieutenant Deacon registered for call up under the new mobilisation law in September 1939 at the age of 22. He was a member of the Croydon Radio and Television Society and, given his preference for the Royal Navy, he volunteered for the Wireless Branch. He received his call up papers in December 1940, and entered *Royal Arthur* at Skegness for new entry training as a steward. There were large numbers of wireless operators also undergoing training there and Lieutenant Deacon put in a request for a transfer. At the time his Divisional Officer carefully explained that 'It had already cost the Crown a lot of money to kit young Deacon out as a Steward, and if everyone wanted to transfer we would not win the war…' However, in February 1941 he was ordered to present himself at HM Signal School Portsmouth for interview. When asked what he knew about RDF (the original designation for radar), Steward Deacon observed that 'DF' was ordinary wireless 'direction finding' and he then deduced that the 'R' must refer to 'ranging'. The interviewer dropped his monocle and remarked to the Commander Chairman 'If this man knows the meaning – does the enemy?' Two days later Wireless Mechanic Deacon joined eleven others to start their radar training using the operational radar Type 279 at the Royal Marines barracks in Eastney, which formed part of the Southern Command Early Warning System. Apparently, the most senior radar man on the course was John Brean, who had previously been chief engineer to John Logie Baird, the accredited inventor of television.

Lieutenant Deacon then went to sea in HMS *King George V* with a roving commission around the Fleet to fix radar problems. In particular he found himself providing support to the destroyers fitted with ex-RAF ASV fixed array radars (Type 286). These radars were initially designed for air to surface detection and used a vertical scan on the cathode ray tube. When used from a surface platform, the object was to make a contact show as butterfly wings on the trace, with the ship turning to make the wings equal in size.

In April 1942, a new rating category was created and he was rated Leading Radio Mechanic. At one point he was due to return to his ship in Scapa Flow only to miss it as it had sailed to engage the *Bismarck* leaving him to follow the news of the loss of HMS *Hood* and the sinking of the German battleship. However, he was aboard *King George V* escorting one of the Murmansk PQ convoys in 1942 when HMS *Punjabi* – one of eleven destroyers in the anti-submarine screen – misjudged a screening manoeuvre and was cut in two by the battleship. The stern of the *Punjabi* sank almost immediately and, being full of primed depth charges, exploded and covered the *King*

George V with a shower of fuel. During the collision most of the battleship's radar and radar equipments suffered damage and Leading Radio Mechanic Deacon was sent aloft to repair a gunnery control equipment aerial 140 feet over the water with the ship suffering a twelve degree list.

In September 1942, Petty Officer Deacon appeared in London before a Commissioned Warrant (CW) selection board which included Frederick Brundrett,[121] later to become Sir Frederick Brundrett who in 1946 became Chief of the Royal Naval Scientific Service. During the interview his illustrious examiner appeared to be more interested in quizzing the candidate about the performance of the new Type 273 surface warning radar recently fitted to *King George V* rather than the candidate's promotion potential. Nevertheless, the result was a letter in November awarding him a commission as a Temporary Sub Lieutenant (Sp) RNVR and appointing him to a long radar course in Portsmouth.

In 1944, Lieutenant Deacon also became involved in the loan of the HMS *Royal Sovereign* to the Soviet Navy, a gesture made by Winston Churchill to match the handing over by President Roosevelt of a US cruiser. The *Royal Sovereign* was well fitted out with modern equipment and Lieutenant Deacon was part of a team sent to the ship to remove 'sensitive' equipment prior to handover. USS *Archangelsk* then departed for Russia with a motley crew including boy seamen of only thirteen and fourteen years of age.

Following the cessation of hostilities, Lieutenant Deacon was made the base radar officer in Wooloomooloo, Sydney, Australia, from where he flew to Hong Kong in the van of the Royal Marine task force who were taking the territory back from the Japanese after their capitulation. It was here that he met the first Commander (L) in the Royal Navy, Commander Geoffrey Turner, who encouraged him to apply for a transfer to the Electrical Branch. This he did, but before returning home he cruised the Inland Sea of Japan in HMS *Tyne* and visited Nagasaki and Hiroshima to witness the dreadful effects of the atomic bombs. Eventually he came home on a compassionate draft to visit his sick father, travelling a good part of the 21-day journey in the bomb bay of a US Liberator. With nothing further heard of the transfer request, he was posted for demobilisation and discharged with effect 3 June 1946. Three years later the Korean War began to make demands on the Royal Navy and his name was drawn from the emergency list. This resulted in him eventually joining the branch in March 1949 as a Lieutenant (L) on a five-year short service commission.

Lieutenant Austin Edbrooke RN

Lieutenant Edbrooke joined as an Electrical Artificer Apprentice in 1940 and found himself in the Electrical (Torpedo) School, HMS *Defiance* in 1944. Sent to sea as a 4th Class Electrical Artificer, he was on board the destroyer HMS *Quality* in Tokyo Bay for the Japanese surrender on 2 September 1945 just before the ship was transferred to the Royal Australian Navy in October 1945. After being rated Chief Electrical Artificer in 1951, he was subsequently selected for promotion and joined Class SD11. After successfully completing the course, Lieutenant Edbrooke became one of the first Branch List electrical sub lieutenants to be promoted from the lower deck.

In 1949, the ranks of WO and CWO were changed to 'commissioned officer' and 'senior commissioned officer', the latter ranking with but after the rank of lieutenant, and they were admitted to the wardroom, the WOs messes closing down. Collectively these officers were known as 'branch officers', being re-titled 'special duties' officers in 1956. In 1998, the Special Duties list was merged with the General List of officers in the Royal Navy, all officers now having the same opportunity to reach the highest commissioned ranks.

Commander F.J. Emuss OBE RN

Commander F.J. Emuss was accepted into the RNVR in 1937 as a yachtsman and was duly granted an executive commission when war broke out. His first job was training sailors at HMS

Ganges, Shotley until it was evacuated to a camp site near Gloucester because the Germans were expected to invade that part of the east coast. The threat of invasion was lifted and so the training returned to Shotley. However, for Commander Emuss the return was to be short-lived as it was discovered that he had been a practical radio ham. Without any training he was immediately diverted to fitting RDF to ships. After some 50 ships were complete, he was sent off to Gibraltar with a team of four assistants and a dozen mechanics to service the convoy escorts. He then joined Force H as the Squadron Radar Officer for the North Africa, Sicily and Italy invasions.

Returning to the UK Commander Emuss joined the ANCXF planning staff for the Normandy Invasions with responsibility for radar, RCM, navaids and diversions. After the war he became an electrical officer in the Royal Navy, and saw service in HMS *Dryad*, the Royal Dockyards, London Whitehall, Bath and HMS *Vanguard*. He also served in submarines in the Pacific.

Captain C.P.H. Gibbon RN

Captain Gibbon joined BRNC Dartmouth as a 13-year-old cadet in 1937 with an unusual ambition at the time to become a (Marine) Engineer. However, it was suggested that because he had good eyesight and it was wartime he should put his ambition aside and go to sea as an executive officer. Consequently, as Executive Branch Midshipman Gibbon, he joined the Town Class light cruiser HMS *Manchester* in January 1941. The ship was torpedoed during a Malta convoy in August 1941 by Italian MTBs but made it back to Gibraltar where Gibbon was transferred to the destroyer HMS *Foresight* escorting Force H and Atlantic convoys. After four exciting months, he left *Foresight* to complete his remaining midshipman's time in the battleships HMS *Rodney* and HMS *King George V*.

After completing sub lieutenant courses in the summer of 1943 and submarine training in *Dolphin*, he joined the submarine service where, via HMS/M *Otus* and P511, a First World War ex-US Navy S Class submarine, he joined HMS/M *Vampire*. In autumn 1943 after work up in the Clyde, *Vampire* carried out patrols off Norway before sailing for Malta to join the First Submarine Flotilla. Shortly after joining the flotilla and well short of his 21st birthday, Sub Lieutenant Gibbon found himself as the first lieutenant of the *Vampire* and during this appointment he earned a Mention in Despatches for the *Vampire's* success on patrol even at that late stage of the war. Early in 1945 as the First Lieutenant standing by HMS/M *Auriga* in Barrow, he was about to deploy with the submarine when VJ Day intervened and the submarine remained in the Clyde.

When the formation of the Electrical Branch was announced, despite well-intentioned discouragement from his fellow submariners, Lieutenant Gibbon applied to join. He remembers Captain S/M Clyde telling him that the Electrical Branch would 'Equate with the "Schoolies" and dentists and that there would never be electrical engineers in submarines'!

Although considered too old to go to Cambridge for degree training, Lieutenant Gibbon was thought young enough to benefit from an academic 'top-up'. In autumn 1946, along with some fifteen other would-be electrical officers, he joined RNC Greenwich for twelve months further education for which Captain Gibbon remained profoundly grateful. After Greenwich he was sent to *Collingwood* and joined Class 33L for electrical officer training and radio conversion courses. In the final event he never completed the RCC, but was given a jump back to the Third Submarine Squadron (the term flotilla now applied to the whole submarine command). In December 1950 he was summoned to HMS *Dolphin* by the Chief of Staff, the same Captain S/M who had advised him on his transfer aspirations, who told him with humour and no little irony that he should join HMS/M *Taciturn*, then converting in Chatham, as the first submarine electrical officer. Looking back, Captain Gibbon found it astonishing how unprepared the Royal Navy was to cope with the new technology being presented and he recalls it as 'a slow and painful learning curve for everyone concerned'.

Rear Admiral John S. Grove CB, OBE

Rear Admiral Grove entered the Electrical Branch in a most convoluted manner because on his first attempt he was unsuccessful, having failed the eyesight test for the Navy. This was in spite of sitting the special entry examination in June 1944 and being awarded marks that were higher than several successful applicants. Consequently, he headed north to read electrical engineering at St Andrew's University where he obtained a first class honours degree in 1947. As Mr Grove he was then called up and joined the Mons Officer Cadet Training Unit, Aldershot for initial training as a National Service Technical Officer before joining the Royal Engineers and being sent to lecture in physics at the newly formed Royal Military Academy Sandhurst. Shortly afterwards, he succeeded in transferring to the Royal Navy and joined as Instructor Lieutenant Grove in December 1948. Quite remarkably, considering the reason for his earlier rejection on eyesight grounds, he was sent to teach practical navigation to executive midshipmen onboard HMS *Forth*. While in Malta in 1949–50, he discovered the Electrical Branch, and, with his background in electrical engineering, he wrote to the Admiralty and presented a case which resulted in a second successful Admiralty Interview Board and Lieutenant (L) Grove joining Long Course 2(U2) in October 1950.

Commander Dennis Guy RN

Commander Guy joined the Royal Navy as a sub lieutenant RNVR officer in 1941 and received two weeks of training in *Vernon*, which by then had been relocated to Brighton at Roedean Girls School. He was awarded a commission in the Royal Navy and was sent, as Lieutenant Guy, on Radio Conversion Course (RCC) 1 in *Collingwood* in1946. The first course was six months long but this was increased to nine months for RCC2 following the feedback given by earlier students, who had gained much practical experience during the war but at the expense of their academic development.

Lieutenant Guy was sent as Assistant Electrical Officer in HMS *Birmingham*, where the newly founded Electrical Department had taken over the 110 Volts DC generators from the Engineering Department, the power distribution system and electrical equipment from the Torpedo Department, the wireless equipment from the signals officer and the radar from the RNVR radar officer.

The Electrical Branch was formed by Royal Navy or RNVR torpedo officers, RNVR electrical officers – who were in the majority – and civilian officers who worked in the dockyard electrical department. It is well known that many personnel did not like these major changes, and the electrical warrant officers, CPOs and POs who had served in the Royal Navy for many years were not happy or cooperative about ex-RNVR officers taking charge of them.

While *Birmingham* was in the Far East, the Electrical Officer, a lieutenant commander, was invalided back to the UK. The Captain was undecided whether to let Lieutenant Guy take over the Electrical Department or to hand it over to the Torpedo Officer, but Guy pressed his case for the recognition of electrical specialisation and eventually took over the Department.

Following an appointment to HMS *Harrier*, the aircraft direction centre in Pembrokeshire, testing 960, 982 and 983 radars before installation in ships, Lieutenant Guy joined Captain D's staff in the Northern Ireland as the Electrical Officer of the Londonderry Anti-submarine Training Squadron where he and his deputy looked after four frigates and two destroyers.

His next appointment was as the Drafting Officer for electrical and shipwright personnel in the Chatham Port Division ships. As such, he visited all Chatham-manned ships coming into their base port, and during this time he became convinced then that the Electrical Branch had been accepted by the Fleet as an efficient and effective branch.

There followed an appointment to train CPOs and POs in leadership at HMS *Royal Arthur*, Corsham. Although he was not pleased to be sent to another non-electrical job, he was told it was because there were no sea going jobs available. It was while at Corsham that he recalls that the green stripe between the gold braid was removed along with the purple of the marine engineers and white paymasters' coloured stripes.

Commander Michael Hunter-Jones

Commander Hunter-Jones joined the RNVR as an Electrical Midshipman direct from university in late 1943 and was transferred to the new branch on its inception in 1946. After courses in *Collingwood*, he was sent to the Mediterranean for nearly three years onboard HMS *Cheviot*, during which time he became involved with the Palestinian Campaign. On returning ashore he became one of the first naval officers to be appointed to the Directorate of Electrical Engineering in Bath although he was actually based at the Admiralty Engineering Laboratory at West Drayton. This was during the time when the Royal Navy was changing over from DC to AC ships and Commander Hunter-Jones had the task of evaluating new equipments for the AC ships. With this experience, he was then appointed to Headquarters to write procurement specifications and advise contractors on AC electrical requirements for new equipment.

Commander Hunter-Jones then went to HMS *Vanguard*, where as an electrical officer he claims to have been the last battleship fire control officer in the Royal Navy or, in his own words, 'the last man to control guns that were big enough to be called guns'!

James McKenna RNSS

James 'Mack' McKenna became associated with the electrical systems development for the Royal Navy as member of the Royal Naval Scientific Service in 1945 when he was assessing communications systems and developing installation guidelines for new systems in the Fleet, naval air stations and shore wireless stations, with particular emphasis on HMS *Vanguard* and RNAS *Culdrose*. His other work included the early days of electronic warfare systems development and preparing the ships for the nuclear weapon trials conducted at Christmas Island.

As individual ship equipments started to evolve into more complex and integrated combat systems, Mack McKenna's involvement with installation planning eventually led him to become the first whole ship weapon system engineer for HMS *Bristol* which was the concept ship for the first of the ADAWS systems.

Commander Cyril Locke RN

Commander Cyril Locke started his career in 1939 as a Temporary Commissioned Probationary Sub Lieutenant Special Branch RNVR working with a degaussing team after the discovery of a magnetic mine at Shoeburyness before joining HMS *Furious* in 1941 to gain his bridge watchkeeping ticket. While in *Furious* he sailed with many Malta convoys, including Operation *Pedestal* when nine of the fourteen merchant ships were sunk and two carriers, four cruisers and one destroyer in the escorting force were sunk or damaged by relentless enemy air and sea attacks. *Furious* also took part in Operation *Torch*, the invasion of North Africa.

In 1943, he was sent to *Vernon*, then sited at Roedean Girls School near Brighton, for the Torpedo Long Course. After this he joined HMS *Jackdaw*, a Royal Naval Air Station near St Andrews in Scotland, to teach pilots how to drop aerial torpedoes. As a Lieutenant Commander, Cyril Locke stood by HMS *Vengeance* and sailed with her to the Pacific Fleet and into Hong Kong after VJ day where the crew became involved with the policing ashore.

Having been demobilised from the RNVR in February 1945, Commander Locke restarted his naval career by joining the Royal Navy as a member of the Electrical Branch. His career appointments included HMS *Mercury*, HMS *Gambia*, the radio department at HMS *St Angelo* (Malta) and then into the submarine world at *Dolphin* as Staff Commander (L) to FOSM. Following appointments to HMS *Sheffield*, HMS *President* and the 1st Submarine Squadron at Malta, Commander Locke completed his second naval career as Staff Electrical Officer to Commander-in-Chief Nore at HMS *Pembroke*, Chatham.

Lieutenant Commander Ronald Parkinson RN

Lieutenant Commander Parkinson reports that his early career was a very typical example from that era. On entering the Royal Navy he carried out communications training before being sent to sea in late 1942 in HMS *Unicorn*. In 1944 he transferred branches to become a radio mechanic and was sent for training to Rutherford Technical College, Newcastle. After training he was sent for maintenance duties to HMS *Valkyrie* on the Isle of Man, and subsequently to HMS *Mercury* in Leydene, Hampshire.

LREM Parkinson joined Combined Operations for the Normandy landings as a member of a beach signals unit. In 1946, as a petty officer radio electrician, he entered *Collingwood* for the first time to start conversion training to radio electrical artificer. He served as an artificer until 1955 when he passed the examination for the Special Duties List and was selected for a commission. He was appointed Electrical Sub Lieutenant(R) Officer in HMS *Ceylon* and served on board the ship during the Suez crisis.

APPENDIX 2

GREENIES IN ACTION

The Electrical Branch through to the WE Sub Branch had such a broad technical remit that during wartime the men had a major role in the operation of their ships. The following stories have been included in order to illustrate the breadth of the contribution made by members of the Branch over the years, both during wartime and periods of great change. The author was well supplied with many stories which would have not been out of place in this chapter but, as ever, space has limited the number of accounts.

Walter's War, an Electrical Officer in the Second World War

Lieutenant Commander W.G. Huggett RN

In 1938 and with war imminent, Walter Gower Huggett decided to join the Royal Navy before conscription denied him the choice. Walter had an HNC in electrical engineering and an ONC in mechanical engineering, both of which he had taken at night school while working for his father's engineering company. With this background, as Ordinary Signalman Huggett RNVR, he joined HMS *President*, moored off the Embankment in London, and was subsequently promoted to Signalman Q (Qualified). However as war approached, Walter was sent from HMS *President* to Chatham, thence to Lowestoft and finally Grimsby, ready to join one of the many trawlers due to arrive for conversion for minesweeping duties.

When his ship duly arrived, the last catch of fish was swiftly unloaded and replaced with an oropesa sweep to transform her into HMS *Lune*, pennant number FY588. The fishing crew were found RNR uniforms and Signalman(Q) Huggett was given an Aldis lamp, signal flags, a set of Admiralty Fleet Orders and a sheaf of indecipherable papers. The fishermen were phlegmatic about

Lewis Gun twin mounting. (Naval Museum Portsmouth)

this change of circumstances as they could now look forward to a better watchkeeping system than the three hours of sleep in 24 they had whilst fishing. In addition to his communications duties, Signalman Huggett also had the privilege of being the ship's rum bosun.

The ex-trawlers operated in groups of six under the command of a 'shellback'[122] Royal Navy Lieutenant Commander. Five of the vessels had RNR signalmen who struggled desperately to read Aldis signals being sent amongst themselves from afar at night in the North Sea gales. Fortunately, Walter had a hard-drinking yet heroic skipper who could smell his way round the North Sea and, on one occasion, returned *Lune* safely to port whilst the Royal Navy commander and another trawler went aground in the thick fog.

Every day a convoy would proceed down the North Sea channel and into the Thames, mostly without harassment during the early war days. Occasionally a contact mine would be released by *Lune's* sweep and, as a Bisley contender with the HMS *President* rifle team, it was Walter's task to hit the horns and explode the mine with a 'Ross' rifle of Boer War vintage. Air protection came in the form of a First World War Lewis machine gun, which was supplied without a stand and could therefore not be aimed skywards! As it happened, this limitation was of little consequence as the gun invariably jammed after one round.

Heads Up!

Larger trawlers were soon completing their conversion to minesweepers and eventually Signalman Huggett and his skipper were transferred to HMS *Solomon*, a ship that had a bigger gun but a much narrower bridge access, which caused the skipper at one stage to get stuck in his efforts to get to his post.

In the winter of 1939/40, North Sea Command had a nasty shock when minesweepers from Grimsby, Hull, Harwich, and Great Yarmouth, having swept the convoy channel one day, found that on the next ships were being mined and sunk. A period of near panic ensued until a new type of mine was discovered at Shoeburyness during low tide. It was defused by a mine disposal team and this was the Royal Navy's first encounter with the magnetic mine.

Our wooden fishing trawlers provided too low a magnetic field to trigger these weapons but they were able to detonate off the signatures of the iron cargo ships which followed and that reminds me of the occasion when we were attacked by a long range Dornier aeroplane whilst sweeping the channel. It must have been laying

(Collingwood Magazine Archives)

mines and was not keen on being reported in the area. We survived without casualties but returned to port looking like a colander! Many weird and wonderful devices were suggested and tried out as countermeasures, but it soon boiled down to 'degaussing' where the magnetic profile or signature of each vessel was measured and then nulled by winding coils of wire around the perimeter of the ship through which was passed an electromagnetic field in opposition to the intrinsic magnetism of the ship.

As the signalman I was required to browse through Admiralty Fleet Orders and through doing this I became aware of an appeal for all ranks qualified or experienced in electrical engineering to report this to the authorities. I conformed and thus, in spring 1940, whilst still at sea, I was ordered to report as soon as possible to HMS *Osprey*[123] in civilian clothes. Three days later from being Signalman Huggett I was promoted and given the uniform of a Temporary Acting Sub Lieutenant (Special Branch) RNVR and ten more days to learn all about the operation and maintenance of asdic. I was then sent to Larne Harbour, Northern Ireland, in charge of a non-existent workshop to maintain the asdic fitted to the anti-submarine trawler group and any other vessel that may call.

From peaceful Larne, I monitored the European Blitzkrieg and the Dunkirk evacuation, and became restless enough to volunteer for a transfer, preferably overseas. Being an obvious expert and a volunteer to boot, I was posted to Canada and the United States to learn about American sonar systems. Along with a New Zealand Sub Lieutenant, I took passage in HMS *Richmond*, a converted liner, to Halifax and then Ottawa where we liaised with the British offshoot of *Osprey* who were organising the manufacture of asdic gear in Canada. We then entered neutral US in civilian clothes and visited the submarine signal school in Boston where they made the American sonar, followed by visits to Charleston Carolina and the naval base at San Diego. The Americans were well briefed about our visit and we were given the full treatment. A British-US deal had just been completed which, I believe, had something to do with the strategic use of Bermuda and 40 ancient 'four stack' destroyers,[124] most of which were fitted with sonar. Out on exercise off San Diego against a tame submarine I formed a poor opinion of this American sonar, which could detect the submarine but seemed to be barely capable of tracking it to a successful prosecution. I concede that this was many years ago and the American sonar technology is nowadays quite awesome.

The British asdic of 1940, developed during the inter-war years at Portland, was a clever combination of basic mechanical and electrical principles. The quartz crystal oscillator and its training mechanism was in a streamlined dome that was lowered beneath the hull and which protected the oscillator from the variable wave pressures and turbulence. The American sonar oscillator was housed in a small spherical ball attached to a large diameter shaft designed to withstand the ocean's buffeting. This greatly complicated the oscillator training mechanism and the control circuit was a masterpiece of weird and wonderful electronics. Background noise from the sea however, was never overcome. The American sound tracking system of a flashing light was also inferior to the British chemical recorder, which left a permanent trace on a roll of paper. In fact later, after returning to the UK, I was tasked to 'stitch' the British recorders into the American sonar on board those of the old destroyers that made it to the UK.

Now an international sonar expert, and a tired one after a four-day train ride from San Diego to Halifax, Nova Scotia, I joined HMS *Newport*, one of those ill fated 'four stackers'. Three sailed from Halifax but the *Newport* was the only one to make it across the Atlantic despite being beset with 'condenseritis' and many engine breakdowns. She finally staggered into Loch Foyle, Northern Ireland with fuel for only another 20 minutes of steaming and salt water having to be used to top up the boilers. She finally made it to Devonport for major repairs, was then lent to Norway for convoy duty and terminally expired on her first convoy escort trip to Iceland!

In January 1941, Temporary Acting Lieutenant Huggett was appointed as the Devonport A/S Officer and a member of the new RNVR Electrician Branch, which inevitably was manned by reservists. Plymouth was then the target for intensive bombing by the Luftwaffe and many families deserted the city as night fell, although there were still many casualties from the bombing. In the autumn he left to join a new depot ship, HMS *Hecla*, which immediately sailed for Hvalfjordur, a deep anchorage near Reykjavic, Iceland, in order to service ships and escorts from the Russian convoys. From the relative safety of Hvalfjordur, Walter witnessed new Canadian corvettes arriving from St Lawrence or Newfoundland with 90 per cent of the crew being volunteer landlubbers, and then sailing straight into escort duty for the Russian convoys, where many men and ships were lost in those terrible Arctic waters.

In the summer of 1942, *Hecla* returned to England en route to the Mediterranean but Walter had no chance to go as he was required to remain in Reykjavik because the Port A/S Officer had fallen sick and had been invalided home. However, in the spring of 1943, Walter did return to England and was posted to Harwich as the Port A/S Officer. Whilst this meant mostly servicing the asdic on escorts, he was also responsible for the MTB Base at Felixstowe from which raids were launched over to the continental coast. This role allowed him the opportunity to experiment with hydrophones which the MTBs used to hear the approach of enemy convoys or U-boats. It was a strange coincidence that his father had made hydrophone equipment for a similar purpose in the First World War.

In late 1943, there was another posting, this time one he considered to be a plum job, at the Asdic Experimental Research Establishment (AERE) which, for security reasons, had been relocated from Portland to Fife's Boatyard, Fairlie, Ayrshire. The establishment was run by 'Jock' Anderson with a staff of civilian scientists assisted by a number of naval officers. Walter could not recall any other electrical officers on the staff but there were many civilian electronic experts, some of whom were straight out of university. Even in this celebrated company Walter was not out of his depth and he was able to use his mechanical and electrical engineering experience to develop, amongst other things, an improved control circuit for the motors used to operate the screw lifting gear to lower or raise the asdic domes. The problem had been that while ships used DC power systems, which made it relatively simple to control motor speed, just before the dome reached its lowered operating position the motors had to be stopped just short of the mechanical chocks that sealed the dome into its faired position in the hull. This meant that a sailor then had to go below through water-tight hatches to hand-wind the lifting screw the last few turns onto the chocks.

Walter's solution was to devise a control circuit made from standard GPO relays which were current sensitive. He adjusted the circuit so that the motors were only just turning as the dome reached the chocks, the higher stalling current of the motors was sensed by these relays and the power to the motors was isolated. With his ingenuity the safety of the ship was preserved and no doubt many an AS rating also gave thanks!

U-boats were able to evade early asdic sets by diving deep and this led to a need for a set which could move in the vertical plane, had a narrow beam and operated at a higher frequency, which gave a shorter range but better resolution for depth finding. These were fitted to most of the submarine hunters, including many US Navy vessels. Analysis at AERE suggested that most sets were operating well below their optimum levels of efficiency and were in dire need of tuning. Walter and a civilian scientist, Jim Menter,[125] were dispatched to Liverpool and the ports covering the Western Approaches as the first WESPIT (Weapon Equipment Sonar Performance Improvement Team) long before that acronym was invented. There at the Western Approaches Headquarters they met Admiral Max Horton, who insisted that they expand their influence to cover the US ships fitted with depth detection asdic, both in the Atlantic and in the Pacific for those ships fighting in the war against Japan. Thus on VE Day the team was operating in the Irish Sea, and after a visit to Canada, when VJ Day came around they were working in a US Navy yard in Philadelphia. With victory in Japan assured, work was stopped and a jeep was commandeered to drive to town to celebrate, only to find that the officials had closed all the bars for fear of a riot!

Walter returned to Ayrshire as Lieutenant Commander Huggett and he then heard the news that officers were needed for the newly formed Electrical Branch. Although tempted, he returned to his earlier roots at the Manganese Bronze and Brass Company of Ipswich where he was welcomed back as Technical Sales Manager.

From the L Course to the Yangtze Incident

Lieutenant Commander L.E.D. Wise RN

For any warship to fight in fresh water is rare, but when a 10,000 ton, eight-inch gun cruiser did just that in 1949, the occasion must surely have been unique. Indeed the 'Yangtze Incident' – involving HM Ships *Amethyst*, *Consort*, *London* and *Black Swan* – had several unique features as well as being, I believe, the first time that ships' electrical departments had gone into action as such.

At the time the Electrical Branch was very new indeed and its members had come together from many walks of life. I was in HMS *London* at the end of the war and continued in her as mate of the upper deck during several Far East trooping trips that followed the war, finally putting her into reserve in Chatham in mid-1946. Before this process was fully complete, I was appointed to Royal Naval College Greenwich for a course and found myself one of the 'Terrible Twelve', comprising one Fleet Air Arm observer and eleven salt horse volunteers who were too old to go to Cambridge, too young to have specialised in their original branches and all wanted to be electrical officers.

We formed a unique group known simply as 'The L Course' and I am not sure that the Greenwich staff had any real brief as to what we were to be taught. Roughly speaking, we got something like a three-year university course compressed into one year including maths, physics, mechanics, electrical and radio engineering theory, plus a little bit of chemistry and metallurgy. After years of wartime sea service it was relaxing to get back into the classroom and we were a rowdy lot given to setting booby traps for our lecturers, but they took it in good part for we did work hard too. The subjects were very theoretical but they certainly got our brains working!

Then it was back to reality at HMS *Collingwood*, still all together on Course 33L9 for 33 weeks of general electrics which, with leave periods, took up most of another year. After that should have come the radio conversion course, but the courses were still oversubscribed at that time and half of us had to leave for a breath of sea air, an option which I was, by then, glad to take.

And so it was that I found myself, half trained, in a troopship bound for Hong Kong where I eventually caught up with my new ship, which turned out to be my old ship, *London*. My role as Deputy Electrical Officer was a new one, and I was wearing my green stripe with some pride. It seemed that the ship had not been long in reserve when a requirement arose for a Far East flagship and Their Lordships' eyes fell favourably upon the *London*, long in the tooth maybe but airy and comfortable for flag showing with plenty of accommodation for an admiral and his staff. She was based in Hong Kong, but we often found ourselves on the way to Singapore, sometimes for a docking or refit, sometimes because of troubles in Malaya.

We also made occasional visits to Shanghai and provided support for its considerable British population in those troubled times. Shanghai was still in Nationalist hands, as was the capital Nanking, many miles up the Yangtze River, but the Communist Chinese People's Liberation Army (CPLA) was getting closer all the time. Unfortunately, the CPLA had just established itself on the north-east bank of the Yangtze when HMS *Amethyst* moved up river on her way to Nanking, where she was to take over from HMS *Consort* in supporting our embassy and other British residents. Thinking, I suppose, that this alien warship must be aiding, if not actually allied to, their Nationalist foes, the CPLA artillery opened up at short range and put *Amethyst* out of action and aground on a sandbank.

This was on 20 April 1949 when, coincidentally, *London*, with the Flag Officer Second in Command Far East and his staff on board, was approaching Shanghai where she was due to celebrate St George's Day. After a signal from *Amethyst* had told us the worst, we gave Shanghai a miss and carried on to an anchorage a mile or two further up the Yangtze.

The next stage was for the destroyer *Consort*, on her way from Nanking to Shanghai, to try and tow *Amethyst* clear. Again the soldiers opened fire and scored several hits. With the obvious risk of becoming a second casualty and little room to manoeuvre in the narrow channel, *Consort* had to abandon her rescue attempt and carried on down to berth alongside *London*. I took a team on board *Consort* to assist in damage repairs. This sudden immersion in the unpleasant, sometimes gruesome post-action sights and smells was offputting to say the least when we knew that we ourselves would be going up river in following morning.

The hope was that a much bigger ship, with a flag of truce at the masthead and large Union flags hanging over each side, accompanied by the frigate *Black Swan*, would either frighten or shame the attackers into silence and we would be able to reach the beleaguered *Amethyst* unharmed. Alas, the Chinese artillerymen had not read the rules and we too were soon the sitting target as we steamed steadily up the river. In my somewhat primitive Secondary DCHQ (Damage Control Head Quarters) there was little to do except listen to the occasional 'bong' of a projectile on the armoured hull and wonder what was happening in the outside world. Then a moment of excitement came when I was able to announce on the broadcast that Main DCHQ was out of action and that we had taken over control of the ship's DC. However, the excitement was short-lived, for Main DCHQ's only problem was a blackout after a nearby hit and within a few minutes power had been restored and they were back in control.

There were long periods of silence, but every time we passed a battery we were subjected to relentless close range fire, mostly 40 mm solid projectiles but some shells of 3 or 4 inches calibre which found their mark and most of them exploded on impact. As a result of this unusual opposition there was little or no damage low down in the ship, with the main engines, dynamo and ring main remaining intact, but a great deal of minor damage to the superstructure. Every hit seemed to sever an electric cable of some sort, destroying very many communication and gunnery control circuits. Cables to the director control tower were severed and one turret was put out of action altogether but the others kept on firing in local control. Electrical damage control parties were kept busy rigging emergency lighting and restoring communications where possible.

Sadly it became obvious, as we approached a narrower part of the river and were still 20 miles from our goal, that the hostile reception was set to continue and there would be no chance of returning successfully with *Amethyst* who, although now afloat, could only steam at slow speed. As a final straw, a shell hit the bridge, severely damaging communications, mortally wounding the Navigating Officer, killing the Chinese pilot and wounding the Captain. The Captain then gave the order to turn and handed over navigational control to the emergency conning position from whence the manoeuvre was completed.

Black Swan followed us round as we set off towards Shanghai and, after a few minutes of clearance and repair, command was again restored to the bridge. Fortunately, a Yangtze pilot of great experience, Mr Sudbury, had volunteered to come with us and did a first rate job of getting us home safely. He was later awarded an MBE, as was Lieutenant Commander (L) 'Jock' Strain who had done so much to keep *Amethyst*'s electrical equipment going during her enforced idleness up river.

The return journey was a repeat of the outward one except that we were under fire for longer, the CPLA having brought more guns into action. In all we had been through seven separate actions spread over some three and a half hours. Only as we came into safer waters near Shanghai did we realise that a breaker room had been underwater for some hours without the service fuse release switch opening as it should have done. The Yangtze water had been too fresh to bridge the contacts in the flood switch! Later there was some discussion as to whether a flood switch should have a pot of salt inside its casing to ensure its immediate operation, as some claimed to have been the practice in the past. Others thought it better for the equipment to go on working as long as possible and risk some deterioration of equipment. In the event some corrosion was caused but there was no permanent damage.

Finally we arrived at Shanghai, where we could start to take stock of the situation. Thirteen men had lost their lives and quite a number were wounded (two of whom were to die later). In the Electrical Department we were lucky to get away with one EA slightly wounded and *Black Swan* had no Electrical Department casualties. We could quickly devote our full attention to repair work. We had received over 60 penetrating hits and even the small solid projectiles seemed each to have found an electric wire to sever. It seems from all reports that the new electrical department organisation worked smoothly in all ships and the electrical damage control parties acquitted themselves well.

A further surprise came when the capstan motor was started; water shot out in all directions as it began to rotate and apparently it had been flooded by a fire hose during the action. As the water was fresh, the motor continued to run until the job was done, but it was thought advisable to give the motor and controller a thorough clean-out. An EA dismantled the controller, put the contacts in a bucket of clean water which he left on the deck while he went off to do another job. When he returned, the bucket and contents had gone. Enquiries revealed that a seaman, tidying up his part of ship, had grabbed the bucket and got rid of the 'dirty water' by tipping it over the side! A Chinese fisherman with a net on a long pole was recruited and, remarkably, recovered all the contacts save two, for which replacements were made in the workshop.

Being in charge of low power, I remember poring over ship's drawings to locate damaged cables and climbing around masts and superstructure armed with pliers and insulating tape to make some unprofessional but rapid repairs. After a few days in Shanghai the ship was reasonably fit for action again and we moved to a standby anchorage near the mouth of the river, where we stayed until relieved by HMS *Belfast* on 4 May. Then it was back to Hong Kong for dockyard attention.

Finally, more than three months later, a battered *Amethyst* seized the right moment to make her brilliant and unaided escape down river. *London* by that time had moved south to Malaya and played no further part in the *Amethyst* story. In September, as planned, we returned to Chatham and once more put the ship into reserve, but this time it was for good for she was scrapped a few months later. As for myself, it was back to school once more on the radio conversion course.

The action was unique in many ways, quite apart from the fresh water aspect, the nature of the weapons used against us and the type of damage they caused. For one thing, we kept up a speed of 25 knots throughout, a speed unlikely to have been exceeded in the Yangtze except by *Consort* who had achieved 30 knots the day before. In addition to navigational problems, we were further restricted by the policy of not opening fire until fired upon and this applied to each separate engagement as we passed each shore battery.

After the incident, there was a rare if not unique 'cross-decoration' when the crew of the RAF Sunderland, which had brought replenishments and personnel to the *Amethyst*, were awarded the Naval General Service Medal with Yangtze Clasp, while the RAF doctor who transferred to the *Amethyst* was amongst those awarded the DSC. Finally, I am sure that *London* must be the only ship to have gone into action with 'illuminate-ship' circuits rigged. These had been got ready for St George's Day and, in the event, came in handy for temporary lighting.

Electrical Damage Control in HMS *London* During the Yangtze Incident

Lieutenant Commander E.H. Johnson RN

This section is taken from an article published in the *Naval Radio and Electrical Review* in October 1949. It was concerned chiefly with the action as it affected HMS *London,* for although the smaller ships were equally engaged, the type of armament used against our ship was more akin to what might be expected of a close action between light craft. It was very different from the kind of thing an eight-inch cruiser might expect.

The Chinese People's Liberation Army used batteries of field guns mounted on the flood banks of the river, and mobile anti-tank guns farther back and hidden from view amongst the trees lining the paddy fields. Damage was consequently not great for each incident, but the total of 66 hits produced a big problem for the repair parties. Only thirteen hits failed to create electrical damage and, as most of these were those that exploded against the side, practically everything that came inboard made work for the Electrical Department. Altogether seven actions were fought and in each case the fire was so hot that repair parties could do little but take what cover could be found on the disengaged side, ready to take remedial action during the lulls. The work was nearly all in running emergency leads, as the ring main was untouched and no main supplies

HMS *London* post–1947 refit. (Naval Museum Portsmouth)

interrupted, so the switchboard had a fairly easy time. All four generators were on load, the ring main being run in sections, and individual loads varied between 450 and 700 amperes, Nos. 1 and 2 generators each carrying approximately 200 amperes more than Nos. 3 and 4, as Y Turret was not in action.

One of the early hits partially severed the supply cables to A Turret pump. The pump continued to run for some time and did not fail until the fifth action. Emergency cables were run to the permanent risers from the pump, but as soon as power was restored one of the risers blew out, causing a heavy overload on Number 1 generator, which fortunately held on as the fault cleared itself. The damaged riser was repaired and the pump run via the emergency leads.

The flooding of Number 4 breaker space produced some strange effects in 11D group of breakers. The flood switch failed to open the fuse release switch and the breakers remained alive, continuing to function underwater. There was no indication of the flooding shown at the main switchboard as the earth lamps were already indicating the many earth faults which other hits had produced. Investigation of the compartments in the vicinity of the hit proved that the breaker space was flooded, but the breakers were left to function until dimming of the indicator lamps on the main switchboard after about one hour showed that failure was imminent. At this stage all breakers were taken off and alternative supplies made good. The cross feeding fuses in the section changeover switch of groups 1D also were withdrawn to prevent No. 1 Section being affected through the cross-feeders.

The reason of the failure of the flood switch was not appreciated during the heat of the moment, but subsequent investigation and more leisurely consideration provided the explanation. Not only was the flood water fresh, but it was loaded with a greasy mud and covered with a film of oil washed from the gunner's store, which had received the initial penetration. This had the effect of coating all contact surfaces with a film of insulating material that effectively prevented the flow of enough current to blow the fuses of the fuse release switch and limited the leakage currents in the rest of the gear to harmless proportions.

Considerable numbers of high and low power cables were cut by splinters, causing numerous failures to power, lighting, communications, and fire control systems. This caused the loss of all communications from the forward director control tower firing circuits, to the port 4 director and much of the bridge communications, and caused the loss of the majority of permanent lighting in the forepart of the ship. The port crane suffered a direct hit in the main hoist controller and also damage to the structure, notably the teeth of the training rack, which caused some wardroom argument on who should do the repairs, a strong case being made for Toothy [the dentist].

Over 800 lamps were damaged or failed during the action. Some W/T gear suffered temporarily as aerials were shot away, but only Type 89 and Type 49 equipments received superficial damage by blast and splinters when a compartment adjacent to No. 1 transmitter room was hit.

Of the radar equipment, only the Type 277 remained in service while the remaining sets were undamaged below decks except the Type 79 transmitter, which had an anti-tank shell through the power board where several leads had been cut. The rest were out of action only through damage to aerial and power supply leads, with the exception of Type 284 aerial, which was destroyed by a direct hit. After the action, the heavy work for the Electrical Department began when emergency repairs were taken in hand. As we were not at war and not expecting any more fighting, conditions of navigability and comfort were given priority over gunnery and work was begun to this effect. The quantity of black tape used was prodigious and by the third day after the action we were a going concern with well lit mess decks, adequate ventilation, bridge communications in order and wireless aerials restored. In addition, sufficient work had been done on gunnery circuits to enable the main armament to be fired in director control.

Subsequently, difficulty was experienced with the capstan engines which had been flooded by enthusiastic fire parties dealing with a burning canvas store, but they held together well enough to get us back to Hong Kong where they were handed over thankfully to the base support organisation.

The department organisation worked well, and, with one exception when a telephone order was misinterpreted, communications between sections and the Damage Control Head Quarters were passed efficiently and acted upon with despatch.

In *London* and HMS *Black Swan*, the Branch was fortunate to suffer only one slight casualty, but in HMS *Consort* and HMS *Amethyst* we took some sad losses. On her return to Hong Kong, *Consort* was seen to be wearing a mysterious long blue pennant. Enquiry elicited the answer that it was the Blue Riband of the Yangtze, self-awarded for the passage of this very difficult river at the somewhat unusual speed of 30 knots!

Operations *Grapple* X and Y, the Electrical Branch on Christmas Island 1957–58

Engineer Lieutenant (RE) D. Fordham RN

Dave Fordham was on a radio mechanician's course when he was given a pier head[126] jump to go to Christmas Island as part of the Base Support Organisation for the British nuclear testing programme in the Pacific.

Testing of Britain's first nuclear fission weapon, the atomic bomb, was carried out in the Monte Bello Islands off the west coast of Australia in October 1952. The test was a simulated harbour

HMNZS *Rotoiti*, ex-HMS *Loch Katrine*. (Collingwood Museum)

HMNZS *Pukaki*, ex-HMS *Loch Achanalt*. (Collingwood Museum)

explosion with the weapon being detonated on board HMS *Plym* and, as the controlling Task Force was largely seaborne, the Royal Navy was in command of the operation. Other atomic weapon tests took place later at the Maralinga test site in South Australia.

When later in the fifties the first nuclear fusion weapon, the hydrogen bomb, was developed, greatly increased yields were expected and this dictated that the tests be conducted at a rather more remote location. The Line Island Group in the Central Pacific was chosen. Since these were to be largely air dropped weapons, the Task Force was commanded by the RAF and the main base was established on Christmas Island with the target island being the uninhabited Malden Island 400 miles south. Thus Operation *Grapple* was set up with the control and monitoring being carried out at sea owing to the small size of Maiden Island, roughly a triangle about 5 miles along each side. A small Task Force was assembled consisting of HMS *Warrior*, the tank landing ships HMS *Narvik* and HMS *Messina*, and two Loch Class frigates of the Royal New Zealand Navy, HMNZS *Pukaki* and HMNZS *Rotoiti*. *Narvik*, a veteran of the Monte Bello atomic tests, was fitted out for scientific technical control and monitoring and the first three British fusion weapons were tested at Malden Island in early 1957.

The Operation *Grapple* tests and the data obtained made it clear that it would be safe to conduct further tests on Christmas Island itself. This eliminated the logistical problems of a test site 400 miles from the main base, with the attendant accommodation difficulties and the problem of access to and from the island to set up and monitor the tests. It was decided, therefore, to greatly expand the supporting task force and base facilities on Christmas Island and establish, with the Cold War at its coldest, a permanent nuclear testing site.

In the summer of 1957 a shore-based naval task group was set up and, as Radio Mechanician Fordham, I was pulled from the last few weeks of my qualifying radio mechanician course and sent out to the Pacific on something of a pier head jump!

After the completion of *Grapple* and the decision to mount all further tests at Christmas Island, the build-up began in earnest for what became known as *Grapple* X, Y and Z. The naval group, called rather whimsically Naval Party 2512, was quite small and besides the Resident Naval

Officer boasted only seven other officers – a first lieutenant, two marine engineer officers, a supply officer, a shipwright officer, a boatswain and a Captain of Royal Marines in charge of a landing craft squadron. There was no electrical officer, the electrical complement being Electrical Mechanician Sam Weller, one radio electrical mechanician (myself), two electrical mechanics (EMs Tierney and Harrison) and one radio mechanic (REM Brown).

The chief task of the electrical staff, which seemed a little obscure at first, was to set up a maintenance base for the landing craft, small boats and motor transport. On the radio side, a large number of portable wireless outfits, Type 62 sets and Type 88 sets, were in use to provide communication with the landing craft, ships at the anchorage and with an MFV used to catch fish for monitoring purposes after each test and as a supplement to the menu. A Type 691 UHF outfit and a TCS were installed later and assistance provided to passing ships on a demand basis, chiefly to the RFAs but also on occasion to warships on deployment. A cinema outfit of two 16mm projectors was also maintained on an open air cinema site and a substantial film library was held to provide exchange films to both warships and RFAs.

I arrived on the island in mid September 1957 and, with the next hydrogen bomb test codenamed *Grapple* X due in November only two months away, life became a little hectic. A full seven days a week routine, 0800 to 1800, was instituted which was to continue until well into 1958 when we dropped to and remained on a six-day week.

All the accommodation and workshops in mid 1957 were in tents, with the exception of the main galley and the Resident Naval Officer's (RNO) office which was a 10 foot by 8 foot garden shed! Power for battery charging was provided by two small petrol generators at first but these were augmented by two large antiquated single cylinder diesel generators as more equipment began to arrive.

Main AC power throughout the island was provided by the Army. The supply cables strung along the tent tops deteriorated badly in the heat and humidity and resulted in two fires, one in the senior rates tent lines and the other in the electrical workshop. Unfortunately, the salt water used to douse the latter contaminated all the batteries on charge at the time, which engendered a great deal of recovery work.

HMS *Warrior* electrical department. (Collingwood Magazine Archives)

Test view from HMS *Warrior's* bridge. (Collingwood Magazine Archives)

The *Grapple* X test in November 1957 found all non-essential personnel on the island assembled on the sand spit at the port at first light and the Gilbertese villagers were all shipped out to HMS *Messina* to be shown films below decks until the test was complete. We were dressed in No.8 AWD (Action Working Dress) with anti-flash gear and required to sit with our backs towards ground zero 28 miles away near the south-east corner of the island, our fists knuckled into our eyes. After a countdown of 20 the flash of the burst was clearly seen through fists and eyelids. At the same time we felt a distinct warm sensation on our backs as if a 1kW electric fire had been held about a metre away for a couple of seconds. There was then a count up to ten after the burst before we were allowed to stand and turn to view the result. By this time the sun was well above the horizon to the southeast and at first that was all we could see, or so we thought, but after a while this sun began to grow larger and fade and the real sun became visible nearby and was at first pale by comparison. The fireball of the bomb rose and dulled and the familiar mushroom cloud with its skirts of condensation falling down the stalk developed until it towered above us seeming to fill the south eastern quadrant of the sky. Two and a half minutes later we saw a ripple crossing the lagoon towards us and we heard a rather disappointing blast, a double report not dissimilar to the now familiar supersonic bangs and of about the same intensity. We then had an unprecedented make and mend for the rest of the day.

On Christmas Eve 1957, the 180th anniversary of the island's discovery by Captain James Cook, the Naval Party 2512 designator was dispensed with and the base was commissioned as HMS *Resolution*. This was the one occasion during our stay when a formal divisions was held – but we did not go over the top and hold a march past!

In April 1958, we again assembled on the sand spit for the *Grapple* Y test and the routine was much as before except that this time we were all equipped with white anti-radiation overalls with hoods, presumably because *Grapple* Y was to be of greater yield than *Grapple* X. This protective clothing did reduce the heat sensation on the back from that which we had experienced on the burst of *Grapple* X but the visible appearance of the burst was not greatly different, nor was the base wave which followed. Later in 1958, *Grapple* Z was completed using two air dropped devices and two devices held from captive balloons; but I had left the island in September. Towards the end of 1958, political decisions were made to discontinue nuclear testing at Christmas Island and there was a rundown to a care and maintenance situation by the end of the decade.

In 1962, the political wind shifted once again and the Christmas Island bases were reactivated with a joint British and US force of some 3500. A further 24 tests were carried out but by 1964 the rundown was completed and on 29 June that year the white ensign was lowered for the last time on *Resolution*.

APPENDIX 3

GREENIE RECOLLECTIONS

This chapter recalls some iconic recollections of the Electrical Branch. It contains some articles selected by the author from the formative years of the branch as well as a number of personal stories contributed by Greenies who took the trouble to put pen to paper and record their memories for posterity. Again, space precludes the inclusion of all the contributions offered but the following have been selected as representative of the branch ethos, humour and camaraderie which built up quickly from the time of its inception.

Executive Concerns

The fledgling nature of the Electrical Branch was clearly a concern to some senior officers and Commander Cyril Lock tells of the following instruction arising from a letter written by Admiral Mountbatten to Vice Admiral Mansergh, who was chairman of the Officers Career Structure Committee. It read:

The Fleet Electrical Officer is to inform all electrical officers as follows:

The first requirement of every officer is to command, lead and direct the work of his men. Any normal routine work must be carried out by ratings under the general direction of their officers. The only circumstances when an officer should do manual work himself are:

1) To show a rating how to do a job when they clearly do not know how to.
2) To help those ratings when they are up against a problem which they are trying to resolve but cannot.
3) When working on special or experimental equipment.
4) To provide personal leadership to his ratings by working with them on emergency work (e.g. when working all night to get a radar set ready for the next morning).

... AND ANYWAY ANGELA I'D GET MY HANDS DIRTY.

(Collingwood Magazine Archives)

It is normally indefensible for an officer to do manual labour, which is normally the responsibility of ratings, simply because he thinks he can do it better or quicker himself, except in a case of real emergency. To do so simply degrades the authority of the senior rating and in the long run lowers the status of the officer.

The REME

Lieutenant Commander Peter Bates RN remembers that in 1942 the Royal Electrical and Mechanical Engineers (REME) was formed by the Army to meet its increasingly complex technical requirements. The establishment of the REME was underpinned by an Engineering Cadetship Scheme whereby courses of 21 months were set up at various technical colleges throughout the UK. The first courses started on 1 February 1943 with some 4000 places being filled by aspiring REME recruits on courses which had attracted considerable prestige at the time and they were subsequently accredited by the Institutions of both the Electrical and Mechanical Engineers.

The first courses were completed in November 1944 when the war was reaching its final phase. The prospects of demobilisation started to loom and the REME decided to take on only about 10 per cent of the graduating students. The remainder were given the option of choosing another regiment in the Army, or opting for the Royal Navy or Royal Air Force. About 70 volunteered for the Royal Navy as Electrical Mechanics 5th class and they were enlisted in March 1945. They were sent to HMS *Royal Arthur*, a training camp established by the War Department by the commandeering of a Butlins Holiday Camp at Skegness on the east coast. The volunteers were then moved on to HMS *Glendower*, another camp site but this time built by Butlins Holiday Camps at Pwllhelli at the request of the War Department. Here the recruits were given a 13-week course in basic seamanship before moving on to *Marlborough* – previously Eastbourne College – for electrical training on a battleship DC ring main installed in a building that was formerly a garage.

During 1946, when applications were invited for permanent commissions in the Electrical Branch of the Royal Navy, many of the REME engineering cadets still on course responded and joined the Royal Navy as officers.

Semi-official Naval Electrical Publications

During the Second World War the fitting of radar equipment was proceeding at such a pace that it was decided to publish information on developments in the field through a quarterly bulletin sponsored by the Admiralty Signal Establishment (ASE). In March 1944, the *ASE Bulletin* was first produced and it contained a variety of articles ranging from radar theory through servicing and modification proposals to feedback from the Fleet, all mixed in with humorous anecdotes and cartoons in a style which was to be a feature of future similar magazines. A number of works by the *Bulletin*'s resident cartoonist, 'Dink', have been included in this book as they capture the ethos of that time. No other information other than the pen name of 'Dink' has been found. The *Bulletin* continued to be the informal link between ASE and the Fleet until September 1945 when publication was stopped at the end of hostilities.

In 1947, DNLD and DRE decided that there was a need to re-establish this informal magazine link and get the technology message across on as broad a front as possible. In March 1947, this led to the publication of the *Naval Radio Review*, a publication filled with interesting articles, which ranged from glimpses into the future to more pragmatic issues relating to Electrical Branch management matters and maintenance practice. Unfortunately, the first issue was classified as a Confidential Book, which did not help achieve a wide circulation, but this was changed from the second issue when it became a 'Restricted' publication written in a magazine style. Initially, the *Naval Radio Review* only covered radio and radar technology, even though it was sponsored by

(Collingwood
Magazine
Archives)

DNLD, but from October 1948 it was expanded to include electrical subjects in order to widen the appeal to the Electrical Branch as a whole. As such the magazine was re-titled the *Naval Radio and Electrical Review*. The format proved successful in not only disseminating information about new equipment and principles to the Fleet but also getting feedback from the Fleet in a less formal manner than the prevailing official channels, which at the time were too cumbersome to deal with the numerous practical and design issues arising from the new technology.

In 1949, NLD approved the idea of an unclassified Electrical Branch magazine which would be produced by *Collingwood*, as the lead school, with contributions from *Defiance*, *Ariel* and the Chatham Electrical School as well as the Fleet. The magazine was to be funded through advertising, rather than public funds, and sold for one shilling. The aspiration was that the magazine content should be representative of the Branch as a whole and contribute to its development as a coherent entity. Thus in April 1949, *Live Wire* was first published with a broad editorial mix which covered the historical through the topical to the irreverent, all included with the aim of fostering a Branch that was professional, inclusive and had a sense of humour.

Continuing the informal nature of such magazines, one *Live Wire* feature was 'Chokker Sparks' whose antics as a put upon EM were portrayed in a cartoon strip by a resident cartoonist called 'Sessions'. In earlier times these cartoons might have been considered dissident, but given that it was just after the war it is understandable that officers and men who had lived together through so much danger should have a relationship which recognised the balance between respect and servility.

Live Wire continued to be published three times a year until Christmas 1958 when the last issue was produced. The editor at the time was Lieutenant Commander V.C. Dunne and he was able to persuade NLD that some features of the *Live Wire* should be incorporated in the *Naval Radio and Electrical Review* of which he was also the editor. In July 1956, the name was again changed to the *Naval Electrical Review* to bring it into line with the fact that 'electrical' was the collective term used to refer to all radio and general electrical activities. In July 1960, following the demise of *Live Wire*, more general articles associated with the Electrical Branch started to be introduced. These included details of senior electrical officer appointments, personal biographies of flag officers as well as news items from the *Collingwood* and *Ariel* electrical schools.

"CHOKKER SPARKS"

(Collingwood Magazine Archives)

(Collingwood Magazine Archives)

In April 1986, the publication was re-titled as the *Review of Naval Engineering* following the implementation of Engineering Branch development and in recognition of the Marine Engineering Branch assuming responsibility for matters electrical. It was around this time that 'Tugg' started to grace the pages with his insightful cartoons. The *Review of Naval Engineering* continued to be published four times a year and fulfilled the same basic function as its predecessors of getting engineering news and views throughout the Engineering Branches. In 2010, the magazine was combined with the *Journal of Naval Engineering* under the title *The Naval Engineer*.

Electrics – Torpedo School Style

Commander Roy Bigden recalls that at the time when the new Electrical Branch was formed, he was serving in HMS *Indomitable* as a very new Acting Warrant Electrician. The ship was acting as a troop ship at the time and it had a small ship's company augmented by officers and men awaiting release from the Service. On arrival in Portsmouth, some of the STs, who had been carrying out electrical duties, departed to join the new TAS Branch and were replaced by ex-STs who were now re-categorised as LEMs and EMs, most of whom were awaiting release. One of the LEMs had no electrical experience at all and was incapable of working on electrical maintenance and repair. However, with the grudging agreement of the Chief EA, who had given him a careful briefing on his duties and responsibilities, the new LEM was made a senior watch keeper. The ship duly sailed for Australia and, while crossing the Bay of Biscay, Warrant Electrician Bigden received an urgent request to meet the Chief EA in the electrical office. The Chief EA then reported that, as was his custom at sea, he had gone to the main switchboard and enquired of the senior watch keeper, the LEM, the state of the board. 'Fine, thank you Chief' came the reply. 'No' said the Chief, 'What I want you to tell me is what generators are on and how the board is split'. This completely perplexed the LEM and for a while he gazed at the display for inspiration until a hopeful smile lit his face. 'Well Chief – number 3 is on shore supply!'

Sonar Training in the US Navy

Commander L.P. Frith believed that he was the only RNVR officer to be at sea with the US Navy in the Pacific when the Japanese attacked Pearl Harbor. This situation arose after he was sent from the RN on temporary loan to the Royal Canadian Navy and then to the US Navy to learn about the sonar installed in 50 of the 'four stacker' First World War destroyers destined for North Atlantic convoy duties.

> The executive officer of the US 11th Naval District, San Diego, Lieutenant Commander Kinkaid, ordered me to wear the yellow Pearl Harbor ribbon despite the existing British ruling that no other national medals should be worn with Royal Navy uniform. When I returned to the RCN in Newfoundland there was no problem, however, on my return to the United Kingdom I was immediately ordered to remove the medal by the Drafting Commander. Imagine my surprise ten years later, when I was being introduced to the US Navy opposite numbers on joining the British Joint Staff in Washington only to meet, now, Admiral Kincaid. His first words to me were 'Where the hell's your Pearl Harbor ribbon!'

Apprentice Entrance Exam 1943

In an effort to increase the number of artificer apprentices coming into the Service, AFO 6039/43 announced that there was to be a new apprentice entrance exam which was to be an open competition exam, rather than a limited exam aimed at boys from places such as the Royal Hospital School, Holbrook. The cost of entry was set at five shillings, with two exams being held in March and September, with the March exam open to all comers but the September exam showing preference to the sons of Service personnel, killed or invalided while on war duty or who were giving 'zealous and faithful service'. Candidates were expected to have a good standard of education with an unusually specific point of guidance in the announcing AFO being that they were expected to know that '/' meant 'divided by'. Apart from education, the AFO also offered guidance on dental health requirements which mandated that 'the teeth must be in good condition and afford an efficient mastication area, including functionally opposed molars and incisors both sides of the mouth. The loss of five teeth or more will generally disqualify a candidate.' The starting rate of pay was 9d per day, seven days per week rising to 2 shillings per day in the fourth year of the apprenticeship.

Life in Collingwood

Commander Eric Marshall was one of the first electrical trainees to join *Collingwood* in the autumn of 1947 having completed initial entry training at *Excalibur* at Stoke-on-Trent. He remembers that on joining:

> *Collingwood* was just in the throes of changing over from the wartime task of training seamen to become an electrical school. Sea boats and seaplane cranes were to be seen on the huge parade ground, flag staffs were to be seen at the end of the parade ground and on the quarterdeck and the drill sheds were for divisions on wet days. There was, however, already a forest of aerials around the 'White City' and the small parade ground. All accommodation was in wooden huts designed to last just ten years.
>
> In those early days, Saturday mornings were devoted to 'clean ship' with the hands being detailed off on the parade ground for work around the establishment at the behest of the First Lieutenant and the Buffer, to sweep roads, cut grass, tend the flower beds or paint whatever needed (or sometimes didn't need) painting. Special parties fell in on the right and on reaching them the Buffer would ask, 'Who are you?' — 'Pig swill party Chief!' or 'Quarterdeck polishing party Chief!' and so on until at the far right the Buffer and the First Lieutenant came to PREM Parsons and his mate, PREM Mullins, (the 'P' for probationary in those days). 'And who are you then?' 'Measuring party, Chief!' came the quick response; 'Carry on' came the order. The Buffer was quite sure the First Lieutenant knew what the Measuring party was and the First Lieutenant was quite sure the Buffer knew its purpose.
>
> The only ones who were really clear were Parsons and Mullins, and they knew it was a skiving party no less. These two, with 'Lofty' Parsons in the lead as he was an ex-Sea Scout Leader and full of confidence, then proceeded to hold up traffic, have free access to forbidden areas, stop parade ground drills, interrupt training, while they continued to 'measure' the roads, the paths, the parade grounds, the lawns with their piece of string and two sticks, a pencil and a note book, all most formal in appearance. This charade lasted about nine weeks until they made the mistake of trying to measure the road and verge outside the Commander's office. The Commander, a salt horse at that time, watched these two 'artful dodgers' for a while and then, with cane under his arm, scrambled egg firmly on his head and a stern frown on his face, came out to enquire, 'Tell me now, what is the length of your string?' This ended the charade once and for all, for the piece of string was just a 'unit of measure' which had kept these two worthies out of hard work for almost a term. Now their self-esteem and confidence collapsed, but the Commander saw the funny side of it and appreciated their initiative.

The Royal Naval Electricians' Association

Such was the significance of the creation of the Electrical Branch in 1949, that the Torpedo Gunner's Mates Association, which had been formed in 1926, changed its name to the Royal

Naval Electricians' Association. The reason for this was the demise of the TGM rate and the fact that most of the existing members had transferred over to become electricians. In changing its name the Association opened its doors to all senior rate electrical and radio electrical personnel including wartime mechanics and a few telegraphists. The RNEA was registered as a Friendly Society and fully approved by the Admiralty. It was open to all chief and regular electricians in the Portsmouth and Chatham port areas at a subscription cost of twelve shillings per year, reducing to half a crown after payment of ten years annual subscription. Death benefits were £5 rising to £10 depending on length of membership. The Association was instrumental in getting electricians training recognised by the Electrical Trades Union (ETU). While serving, members of the RNEA were given the title of Auxiliary Members of the ETU and full membership on leaving the service if they had previously had six months' membership of the RNEA. A sister association also existed for Devonport electricians. (Taken from *Live Wire*, Easter 1949)

Explosives and Electrical Safety Awareness

John Parkinson, from Wannerooo in Australia, was browsing the internet and spotted information about the *Collingwood* museum and the electronic equipment of yesteryear on display there.

Reading about the equipment displayed in the museum made me sit up, because it was all state-of-the-art to me when I first entered *Collingwood* in August 1947 at the age of 18. *Collingwood* at that time was a huge accommodation area of wooden huts with a few brick buildings in the training area and an enormous parade ground where 2000 men assembled each morning for some physical jerks and jogging before they marched past and dispersed to classes for the rest of the day. The sleeping accommodation consisted of rows of huts, all joined by a corridor and all raised on posts about one metre high. Underneath these huts lived an enormous population of feral cats which kept everyone awake at night by their mating calls and shrieks. Every six months or so, the duty watch would be handed sacks and given the job of catching as many as possible. There would be many a scratched arm before the sacks were filled with a snarling, biting and scratching mob of cats, which were then despatched to be put down.

In late 1947, I was walking into the old wooden *Collingwood* canteen one evening when there was an enormous explosion and all the canteen doors folded and the windows shook. Everyone ran outside to witness a huge mushroom cloud beyond the main gate. The Cold War was in full swing by then and we thought World War Three had begun. Fortunately it wasn't quite as bad as that, but bad enough as part of the ammunition dump on the Gosport Road had blown up. A railway truck loaded with explosives near the wharf had gone up and blown out all the windows across the harbour in Portchester, although the local houses were protected by blast deflectors. As one of the duty watch that night, I spent a miserable four hours in the pouring rain sitting on a pile of 16-inch shells, shaking with cold (or was it fear) keeping guard. Never did we find out the cause of the explosion, but along with the flavour of the time it was put down to 'Reds under the bed'.

I joined in Class 141 and the course we were taking was shortened to nine months of intensive training, with no failures allowed, because the Fleet was screaming out for people to maintain the radar and communication gear. On completion, we were shunted out to ships to maintain a huge range of electronics armed with very little knowledge and no experience of life at sea. One colleague went to a captured German destroyer where he spent most of his time with a dictionary translating the labels on 'his' equipment. I went to the aircraft carrier, HMS *Vengeance,* and, with another 18-year-old, was put in charge of the aircraft warning radars. I think there were about eight people all told looking after all the radar on that ship. There was a Radar Type 79 and a Type 281 which used huge valves for the output at a modest 80 MHz. Radio equipment was handled by the WT Branch who guarded their part of ship very jealously for a few more years before that was finally taken over by the Electrical Branch.

The Royal Navy certainly taught me to be technically versatile and seeing those old PPIs in the museum reminds me that it also taught me to be careful. The first time I sat an exam, I touched my chin on one of the EHT capacitors whilst looking for a fault. The old displays had an EHT of around 3KV but at fairly low impedance. I was unceremoniously thrown off my stool and woke up a few minutes later very shaken but very much wiser. For a safer stunt, the Type 277 radar had a gap in the waveguide just above the transmitter to allow a certain amount of vibration to occur without affecting the loading. It didn't take long for young trainees to discover that a beautiful long arc could be drawn from this gap with one's finger without any burning at all. During ships' open days in the late 1940s this arc was the subject of great amazement and we young Jack-the-lads delighted in demonstrating this phenomena to (female) visitors.

Amalgamated Royal Naval Electrical Artificers' Benevolent Society

The EAs had their own port area benevolent societies and in 1949 they amalgamated and elected a council to represent their views on the future of the Electrical Branch. The Society also offered death and invaliding benefits to its members. Membership subscription was eighteen shillings per year, no doubt to reflect the higher pay scales of the artificer, with death benefits of £30 and invaliding benefits of £20. A novel aspect of the Artificers' Benevolent Society was that on leaving the Service, after paying five years' subscription, the member was entitled to a reversionary bonus of 25 per cent of five shillings per year of membership. Thus after five years' membership a reversionary bonus of six shillings and three pence was due. So that its members did not fall into arrears when at sea, the Society had also arranged with the Portsmouth Trustee Savings Bank to accept allotments from pay and then remit the subscription to the Society on the member's behalf. Meetings were held monthly at the Tramways Hall, Northend, Portsmouth and part of the time was devoted to reading out 'posers', questions sent in from sea by EAs on technical matters, and replying with best advice from the elder statesmen serving ashore. Although not reflected in its title, membership was also open to the REAs. (Taken from *Live Wire*, Easter 1949)

An Early Trade Test?

Lieutenant Commander J.M. Cheverton was under instruction in *Collingwood*.

Early in 1947 I was an Electrical Artificer and was drafted to *Collingwood* for an equipment course on the Mk 37 director and Mk 1B computer, a US weapon control system being fitted to eight Battle Class destroyers. The first months of 1947 were the coldest for a very long time, possibly in living memory. The classrooms in *Collingwood* were heated by gas which came from Gosport but the pressure was so low and fluctuating that the flame in each heater would not stay alight for long. We sat in overcoats and gloves trying to write notes in temperatures near freezing. Each time the flame went out, coal gas leaked into the room and the smell was soon noticed. The person nearest each radiator soon ran out of matches trying to maintain a flame and a vestige of heat. An ingenious device was eventually produced and fitted to each heater; it was a paper clip shaped to fit into and around one of the small gas jets so that once ignition had been achieved the clip glowed red hot. When the gas pressure fell and the flame went out (every minute or so), the clip remained hot enough to re-ignite the gas when the pressure rose again. This simple arrangement obviated the need for the person nearest the heater to continually relight the flame with a match. Nobody knew who the inventor was but he achieved local fame as every classroom was soon fitted with the device!

London Institute City and Guilds in Electrical Engineering

In the early 1920s, instruction for the City and Guilds examination in electrical engineering practice was started in HMS *Vernon*. During the war, those who attended courses at *Vernon* found that they could not complete the course required for City and Guilds accreditation. This created demand for a correspondence course to complete the preparation for the exam and, in 1944, the Admiralty agreed to such a course being made available. The courses were popular amongst the officers and ratings of the Electrical Branch with large numbers availing themselves of the opportunity of improving their electrical theory knowledge. Many ratings realised that City and Guilds certification was of considerable value when seeking employment in civilian life and the apprenticeship did not, at the time, have any civilian recognition. In 1946, management of the correspondence courses was transferred to the *Collingwood* City and Guilds Organisation and the numbers continued to grow. In 1953–54, there were 350 enrolled students and of the five First Class Passes in Electrical Engineering in the country, three were *Collingwood* City and Guilds students. In due course, the EAs apprenticeship was accredited by the British Technical Education Council as a Higher National Diploma and then a Level 4 National Vocational Qualification.

Wartime Fitting of RDF

Commander James Armitage recalled a programme during the war to fit destroyers with a basic RDF equipment for detecting surface vessels which was derived from the AS VII radar equipment used by RAF aircraft. This equipment operated at a frequency of about 214MHz and to get an accurate bearing of the target it was necessary to alter the ship's course to a position where two 'blips'[127] on the linear cathode ray display were the same amplitude. The target was then directly ahead of one's ship. Commissioning this equipment, known as Type 286, in naval ships, again required specialist training and small independent teams of RNVR Special Branch officers were initially deployed to carry out this work. Matching of the aerial feeders was achieved by adjusting their length and this was done by a cumbersome procedure known as the 'cutting back experiment', which involved cutting off short pieces from the ends of the Pyrotenax feeder cables until a match was achieved.

It was not long before base organisations were being set up to support the Type 286 and other types of RDF equipment. In Scotland, a small private hotel, Sherbrooke House in the Pollokshields area of Glasgow, was requisitioned. From being just base support it quickly took on the additional role of a training establishment giving seagoing officers training on the new devices being fitted to their ships and the latest available RDF modifications. Before long Sherbrooke House developed into a bustling 'stone frigate' and much of the credit for this must go to the Officer-in-Charge, Lieutenant Commander O.S. Neill AFC RNVR. Neil persuaded the Admiralty that it was a matter of priority to have a working example of each RDF set installed on the premises to enable training to be carried out and to familiarise ship-fitting staff with what they would be likely to encounter. Rotating aerials made their appearance on the hotel roof whilst motor alternator sets hummed away in the basement. On the domestic side, full use was made of the hotel kitchens, which became known as the galley and a full catering service was instituted on wardroom lines. A 24-hour naval guard was set up and a duty officer roster came into being. Alcoholic refreshment, however, was limited to duty-paid beer drawn from wooden casks.

Mention must be made of the logistics of RDF stores, particularly those required for new installations. The general plan was that the various modules and components should be marshalled from the various manufacturers at a depot in Haslemere from whence they were sent to the shipyards according to a priority list. Initially, not much thought was given to the sequence in which a shipyard required them and frequently a missing item brought the whole installation work to a standstill. A particularly frustrating situation arose because supplies of the resilient mounting on which the main assemblies of Type 285 (600MHz fire control radar) were fitted dried up. It was customary for the fitting out bases to send couriers to collect vitally needed items but in this

(Collingwood Museum)

" ——! WE'VE LOST A —— D6!"

DINK.

case even though the component was simply made from steel pressings and rubber there were no sources of supply and the ship fit could not progress.

A big step forward in logistics was the introduction of the 'plan packing' scheme. The components required for an installation were divided into groups to correspond to the fitting sequence and each group was packed in a crate bearing an identifying letter. Thus, case 'H' would contain framework and mounts, whilst in case 'G' would be found valves and cathode ray tubes. In theory, the scheme should have been the complete answer but inside each case was placed a 'deficiency list' detailing any items which should have been included but they were unavailable, and it was intended to supply the items at the earliest opportunity. Thus, it sometimes happened that a case arrived at its destination less than half full, complete with a list of missing items. There was even an instance on record where the case arrived quite empty with all its intended contents shown on a deficiency list!

In 1940, electronics was new to the scene and many lessons had to be learned the hard way. To cope with the demand for hardware, production rates had to be stepped up to a point where reliability tended to suffer. One quickly learned to diagnose a burnt out transformer by its acrid smell and when a loud bang occurred inside a receiver or display unit it could be assumed to result from a high voltage capacitor exploding. There were many instances where features of design were a sure recipe for erratic behaviour and unreliability. For example, the time base of Type 285 was a plug-in unit with a row of 25 rigid knife blade connectors located in the vicinity of the operator's feet. If one of its seventeen valves became suspect, the unit had to be pulled out, drawer fashion, to fit a replacement and the 25 connectors didn't always re-engage firmly. Because of the location of the time base, it was often found that a sound kick produced the desired result.

In the latter part of the war the development of 10,000 MHz (3cm) components led to the manufacture of a high resolution Type 268 (9400MHz surface warning radar). This proved so successful as a surface search and navigation radar aid that a programme to install it in every warship was instituted. The ship-fitting officers took this in their stride and the radar gave little trouble at sea. The story was a little different from the next 3cm development, the Type 262 (9650MHz close range fire control radar), which was to be fitted to anti-aircraft gun mountings. The radar could scan and lock onto an aircraft target and, with the aid of a tachometric box predictor, gun orders were passed to the gun. The whole of the weapon control system was incorporated in the gun mounting but the RDF modules were in a very exposed position. Although they were enclosed in ventilated steel cabinets, these often had to be opened in adverse weather conditions and the contents suffered accordingly. Unfortunately, the Type 262 was not ready before the cessation of hostilities but it was an important stepping stone towards the tracking radar system.

Commander Armitage also recalls the story of how a destroyer of the Clyde Escort Force acquired the sign of three balls from a pawnbroker's premises and, taking advantage of the secrecy surrounding RDF, fitted the sign in a prominent position on the foremast before letting it be known that it was the aerial of the very latest Type 298 RDF equipment. News used to travel quickly in naval circles and it was not long before the Admiralty learned from several sources that the destroyer was operating an unscheduled RDF set of incredible performance. This caused them to send to the Captain 'D' Greenock a signal on the following lines: 'With what authority and by whom has the Type 298 been fitted to HMS *Nonesuch?*'[128]

What the Greenie didn't tell the Navigator

Lieutenant Dave Fordham tells of when HMS *Unicorn*, in common with many capital ships of her era, was fitted with Type 281 radar.

> Her version, as I remember, was Type 281B [90MHz air warning radar] and it had a manually rotated aerial with flexible feeders that would only allow it to rotate about 190 degrees in each direction. It only had 'A' displays and these were fitted with long persistence tubes. The aerial was a considerable size and rotated by a large steering wheel that would have made an HGV driver feel at home. *Unicorn* was a frequent visitor to Japan during the Korean War and would generally rendezvous with the operational carriers at either Sasebo or Kure with occasional visits to Iwakuni. Kure and Iwakuni lie on the Inland Sea and access was either by the Bungo Suido passage between Kyushu and Shiikoku or via the Shimonoseki Strait between Kyushu and the main island of Honshu. The ship had cause to transit both of these channels many times by late 1951. The Bungo Suido passage was no problem, but it was the longer way round. The Shimonoseki Strait was a short cut but had the hazard of overhead high tension cables, so transit had to be made at the right state of tide to clear the masts and aerials, *Unicorn*'s Type 281 being at the time the highest in the Fleet at 168 feet.

HMS *Unicorn* showing Type 281 aerial still in place. (Naval Museum Portsmouth)

Unfortunately, on one occasion *Unicorn* attempted the short cut via Shimonoseki Strait unaware that new cables had been strung, at least one or more of which were sagging far below their normal catenary, and the inevitable happened. All of the dipole arrays were wiped from the main stem of the radar array and the stem itself bent in the middle at an angle of about 30 degrees. What this did to the Japanese power situation can only be imagined. I was not an eyewitness to this event as I joined the ship on her next trip south to Singapore, but our first job on re-commissioning the ship was to replace the aerial outfit with a new one. Needless to say, whenever we made the transit of the Shimonoseki Strait during the next two years the flight deck was full of 'goofers' prudently stationed at the bow end hoping for a spectacular repeat performance.

Employment Opportunities 1954

Live Wire, Easter 1954:
CEAs and EAs who have completed service time for pensions are invited to apply for employment with Sperry Gyroscope Co. Ltd., Stonehouse, Gloucestershire. These positions are staff appointments. Minimum salary £8 12s. 6d. for a 44 hour week. Pension scheme in operation. Apply to the Personnel Manager, Sperry Gyroscope Co. Ltd., Great West Road, Brentford, Middlesex. Plymouth.

Life of an Apprentice

Ted Stevens was a Series 18 Apprentice under training from 1954–1957. He was asked what it was like to be an apprentice.

It was pretty good. The apprenticeship was meant to be second to none with a lot of studying, a lot of year round sport, and much general enjoyment. As apprentices we felt that we had the upper hand over those in charge of us. At the end of the first year in *Fisgard* we were streamed into the different trades and after a further term split to one of the Part II establishments of *Collingwood* (electrical and radio electrical), *Caledonia* (engine room and ordnance) and Arbroath (air). Coming down on the train to Portsmouth Harbour, the bus from Gosport dropped us in our ignorance at *Collingwood* corner with full kit, for a very long walk. There followed a reception where we were directed to Fisher Section but warned that we must face the OOW's window, salute and wait for it to be returned before setting off on another protracted hike to the Apprentices' Regulating Office and the never-to-be-forgotten RPO Jackson. Nasty, horrible, mean, tall and immaculate, his hobby was collecting Station Cards and picking up apprentices for long hair. Then there was pay day, that fortnightly ritual of pay books on top of caps when you shouted out your number to get a portion of your pay. Only a portion as some was always taken to cover domestics and some held back until leave. This performance would then have to be repeated to get the three tobacco coupons one could exchange for 300 blue liners or pipe tobacco. At six shillings [30p], it presented an enormous temptation to those brought up on surreptitious woodbines.
A great landmark was at the beginning of the sixth term, when one was allowed to move into doeskin uniform as opposed to rough serge. I anticipated the need by getting mine at the end of the fifth term; my foresight was not seen as an attribute and I got 5 days number 9 punishment for jumping the gun. We timed it exactly right for the change from black buttons to gilt, and felt very cool with white silk scarves and kid glove accessories, topped with a GOS [good old soak] cap with the wire grommet removed, and the cap generally reshaped.
Leave was scarce during the instructional term, and long weekends from Friday were certainly never heard of. Saturday noon was the start of the short weekend, but then only alternate weekends as you were on duty for the other. Leave was given one evening in mid week and dog watch instruction occupied all the other evenings. There were Liberty boats assembled in order

(Collingwood Magazine Archives)

to get ashore, followed by an inspection and the senior apprentice marching the others to the main gate. An hour later there was a free gangway, but it still retained the inspection. Not that there was much to go ashore for. Pompey were a top Division 1 team which made Fratton Park one option, but otherwise there were the two cinemas in Fareham or the pubs (the Bugle was a favourite that has stood the test of time). The cinemas competed with the *Collingwood* theatre, which changed its film programme every two days and often featured top live shows.

So it was that the apprenticeship progressed through the workshops, laboratories and classrooms, examinations and test jobs, and final preparation for sea on the weapon equipments, electrics and electronics of the Fleet. A memorable occasion was the final weekend in London for the passing out dinner, the parade and the class being rated up together to 5th Class Artificers.

Fifth Class Artificers went to sea with a killick's hook on their arm, supposedly for a year of supervised probation, yet my first ship was an ocean going tug, HMS *Samsonia*, where I was the senior electrical rating with a staff of one (a JEM). But the terms of probation were presumably satisfied as I picked up my petty officer rate after that year and served for 20 years from when I completed my apprenticeship, finally returning to HMS *Collingwood* on removing the uniform in 1977, where I have served ever since as a civilian instructor in the workshops.

The Gungineer

Captain Howes recalls that while the career path for many after 1946 was complicated and tortuous, he felt that no one had it quite so bad as the (marine) engineer officer who qualified as a (marine) engineer but chose to specialise on completion of the marine engineer qualification course at RNEC Manadon as a gunnery engineer officer. Over the years these individuals underwent the transition from marine engineer to gunnery engineer to ordnance engineer and finally weapons engineer. As such they had some alma mater allegiance to the RNEC Manadon, *Excellent*, *Vernon*, *Mercury* and *Collingwood* but, oddly, not *Sultan*.

Captain Martin Howes obtained his Marine Engineer's watchkeeping ticket onboard HMS *Vanguard*. As an aspiring Ordnance Officer or 'Gungineer', he was required to close up as the safety trainer of the right 15-inch gun in A turret. Firing against a tug-towed battle practice target was routine but it was still made very exciting with the noise of discharge sounding more like an incredibly loud express train passing through a short tunnel rather than a bang. Since the guns on this occasion were to be fired individually rather than as a broadside, it was the erroneous understanding of the safety trainers that their responsibility only lay with their own gun. One gun fired and both safety trainers thought it was their gun. Unfortunately, Captain Howes' gun had not fired and when the gun ran away in training, neither he nor the other safety trainer took any check fire action. The result was a half ton shell dropping only 50 feet from the stem of the tug! These firings proved to be the last made from the ship.

Early Submarine Radar

Captain N.B.M. Clack recalls that through most of the period from early 1963 to the end of 1965, the newly established state of Malaysia was engaged in 'confrontation' with Indonesia. The 7th Submarine Division of five recently streamlined and updated 'A' boats was based in Singapore and involved in a variety of operations on behalf of the fledgling state.

> The submarine radar of this technically early period was the Type 267PW and it was equipped, according to the manufacturer, with a pressure tight seal to its flexible waveguide and aerial within and on top of a telescopic mast. Shortly after operations had begun in earnest, the waveguide seals succumbed to the relatively high tropical ambient temperature and the waveguide filled up with sea water with predictable results. Our urgent requests to MOD (ASWE) for replacements were met with the then customary twin track response. First, confrontation was not war so we were not deemed to be in an operational situation and hence the cost of supply and despatch could not be justified. Second, our attention was drawn to the awesome threats and penalties provided for us in the BR and QRs if any bodge-ups were attempted. There was no reference whatever to the potential of improving operational performance by at least trying for a solution.
>
> Fortuitously, I happened to be in need of fairly substantial dentistry at the time and was whingeing on about the radar problem in the chair when the Surgeon Lieutenant Commander (D) suggested that tooth filler probably had to cope with an environment as severe in its way as the top of a submarine mast! Therefore, why not plug the top of all submarine waveguides with it? The encouraging response of Commander S/M and all his COs was positive and immediate in agreeing to ignore the dire threats in the handbook and one of the leaking seals and its frame was duly conveyed to HMS *Terror*'s dental department where the largest cavity ever presented was most professionally and competently filled.
>
> We had also recently heard of Araldite and so, thinking we knew its comparative transparency to radar wavelengths (one could hardly call it researched knowledge!), we also decided to try and use it as a filler compound. Each of three submarines was fitted with one of the modifications and the trio were given a forenoon to conduct comparative trials, together with the one remaining serviceable original. By midday all boats had signalled that the performance of each of their respective modifications was indistinguishable from the original and the only significant issue might be which of the modifications would prove the most durable. It was a toss-up between tooth filler and Araldite, and as the latter was about a hundredth of the cost of the former, we kept the tooth filler in reserve and so reported to MOD as discreetly as practicable.[129]
>
> The only acknowledgement received was a public one, specifically through a headline in (I think) the *Daily Express* (which had its ways and means) to the effect that one of our submarines

in the Far East had had its toothache cured. Later on the same problem began to affect the Fourth S/M Division in the more temperate waters around Sydney, and we enjoyed the quiet irony of receiving an order to send them detailed manufacturing instructions for both methods by Secret Priority signal.

Weapon Tales

Lieutenant Commander P.H. Marland was on HMS *Blake* during Exercise Springtrain off Gibraltar in 1977.

HMS *Blake* was firing Seacat practice missiles. We had three misfires in a row due to a dodgy batch of igniter primers, each failing on the launcher. The next one lit up but didn't take off. Watching from the console space, I had seen each stage light up, first the gyro flares, then the boost motor followed by the sustainer. I was convinced that the missile had, for some reason, burned out completely and that all we had was an empty casing on the launcher. After the smoke cleared, the PWO then gave the drill order to stow the launcher and thinking it was safe, I readied the console operator to start removing the missile from the launcher after the required safety interval. About a minute later, the supposedly dead missile took off straight up for about 2–300 feet, then like a Tom and Jerry cartoon, turned over and started to fall back down. After the commotion, I understand the Captain and the Navigator ended up getting stuck in the bridge door, as the Captain tried to get out and look at the situation while the Navigator was trying to get in while alternately shouting 'It's falling outboard – no its falling inboard – no outboard!' The reaction of the Sea Cat aimer on seeing the missile was to pull his pram hood down over his head just before the missile splashed about 15–20 yards away. Sometime after the investigations and displeasures were incurred, my section presented me with the catcher net complete with burnt hole, mounted on a trophy board, and you could say that I learned a lot about weapon engineering from that!

While in HMS *Bristol*, as the Ikara Section Officer and Explosives Safety Officer, the Falklands War started and we were ordered to deploy. On the way south, intelligence suggested that if you fired anything in the direction of an Argentinean aeroplane then it went away. WEO came up with the idea of using the Ikara, a torpedo-carrying missile, in an anti-aircraft mode, but the only problem was that it flew at about 1000 feet rather than the 100 feet necessary to give a Skyhawk a severe fright. A select team gathered in the EMR that night, and CPO Carter and I got the books out. The outcome was putting some padding resistors into the circuit controlling the height keeping device in the missile's upper fin so that it nulled out at a lower height. This was tested by putting the whole modified fin into a plastic bag with the Meteorological Officer's barometer and sucking the air out with a vacuum cleaner, whilst comparing millibars and volts. It worked beautifully and the missile would have settled out at about 100 feet. We never got the chance to fire it, but if we had, we would have been able to check whether the vertical dynamics were adequate, because the 55 degree launch angle and powerful boost would have put the missile up to 1000 feet and it would have then come screaming back down, trying to level off at 100 feet, or falling with a very large splash indeed!

APPENDIX 4

CONTRIBUTORS

Armitage, Commander James G.
Bates, Lieutenant Commander Peter
Bigden, Commander R. OBE
Boyce, Commander H. DSC
Cambrook, Commander C.A.
Clack, Captain N.B.M.
Coleman, Lieutenant Commander H.E.
Deacon, Lieutenant David
Edbrooke, Lieutenant
Emuss, Commander F.J. OBE
Fordham, Engineer Lieutenant(RE) D.N.
Frith, Commander L.P.
Gibbon, Captain C.P.H.
Grove, Rear Admiral C.B. OBE
Guy, Commander Dennis
Howes, Captain Martin
Howlett, Commander M.R.C.
Huggett, Lieutenant Commander Walter Gower
Hunter-Jones, Commander Michael
Johnson, Commander (L) E.H.
Langford Lieutenant Commander
Locke, Commander Cyril
Marland, Lieutenant Commander P.H.
Marshall, Commander Eric
McKenna, James 'Mack' RNSS
Parkinson, John (ex-Artificer)
Parkinson, Lieutenant Commander Ronald
Pavey, Lieutenant Commander 'Bill' MBE
Stevens, 'Ted' (ex-Artificer)
Thompson, Commander H.G.S.
Watson, Vice Admiral Sir Philip KBE LVO
Wigney, Captain P.G.
Willis, Rear Admiral Kenneth C.B.
Wise, Lieutenant Commander L.E.D.

WARTIME ELECTRICAL TRAINING ESTABLISHMENTS

HMS *Marlborough* (Eastbourne)	Portsmouth and Chatham Divisions officers and ratings
HMS *Vernon* (Portsmouth)	Junior officers and ratings, all branches
HMS *Defiance* (Devonport)	Devonport Division warrant officers, torpedomen and EAs
HM Torpedo School (Chatham)	Chatham Division ratings, including mine warfare ratings
HMS *Collingwood* (Fareham)	Radio and radar instruction for officers and ratings
HM Signal School (Leydene)	Radio instruction for officers and ratings
HMS *Scotia* (Ayr)	Radio instruction for radio mechanics
HMS *Valkyrie* (Isle of Man)	Radar instruction for ratings
HMS *Kestrel* (Worthy Down)	Electrical instruction for PO air mechanics and ratings
HMS *Kestrel* (Warrington)	Radar and radio training for FAA officers and ratings
HMS *Osprey* (Dunoon)	Electrical instruction (ASW) for all officers and ratings
Sherbrooke House (Glasgow)	Radar instruction for radio mechanics
RAF Training Establishment (Melksham)	Electrical instruction for air mechanics(L)
Chelsea, Aberdeen, Rugby and Walthamstow Technical Colleges	Radar and W/T theory for all air radio mechanics
HMS *Fisgard* (Torpoint)	New entry technical training for EA and REA apprentices
HMS *Caledonia* (Rosyth)	Initial technical and craft training for EA and REA Apprentices
HMS *Vernon* (Portsmouth)	Continuation technical training for EA and REA Apprentices
HMS *Collingwood* (Fareham)	Final technical training for EA and REA apprentices
RN Establishment Arbroath	Workshop and electrical training for air artificer apprentices

BIBLIOGRAPHY

Beanse, Alec *The Brennan Torpedo* (Palmerston Forts Society)

Callaghan, D.N. & C.G. Mount 'The Royal Naval Engineering School HMS *Caledonia Journal of Naval Engineering* Vol 15 June 1965.

Compton-Hall, P.R. 'Two Centuries of Submarines' *Naval Electrical Review* Vol 33 No.2 October 1979

Hackmann, W.D. *Seek and Strike: Sonar, Anti-submarine Warfare and the Royal Navy, 1914–54* (Stationery Office Books)

Harrold, Jane & Richard Porter *Britannia Royal Naval College 1905–2005* (Richard Webb)

Herman, Arthur *To Rule the Waves* (Hodder and Stoughton)

Hill, J.R. (ed) *The Oxford Illustrated History of the Royal Navy* (Oxford University Press)

Hool, Rob 'Development of Naval Minewarfare' (Minewarfare and Clearance Diving Officers Association Website)

Howse, Derek *Radar at Sea* (The Macmillan Press Ltd)

Ireland, Bernard *History of Ships* (Hamlyn Publishing)

Kiely, David G. *Naval Electronic Warfare* (Brassy's Publishing)

Kingsley, F.A. *Radar Equipments for the Royal Navy* (The Macmillan Press Ltd)

Kirby, G.J. 'A History of the Torpedo' *Naval Electrical Review* Vol 26 January and April 1973

Lavery, Brian *Nelson's Navy: The Ships, Men and Organisation* (Conway Maritime Press)

Partridge, Michael *The Royal Naval College Osborne, A History 1903–21* (Royal Naval Museum Publications)

Penn, Geoffrey *HMS Thunderer* (Kenneth Mason Publishing)

Poland, E.N. *The Torpedomen: HMS Vernon's Story, 1872–1968* (Kenneth Mason Publications Ltd)

Preston, Antony *History of the Royal Navy in the 20th Century* (Hamlyn Publishing)

Rawlinson, J.D.S. 'Development of Radar for the Royal Navy 1935–1945' *Naval Electrical Review* Vol 29 No.1 July 1975

Scott, Admiral Sir Percy *Fifty Years in the Royal Navy* (self published)

Smith, P.C. *Ship Strike* (Airlife Publishing Limited)

Wells, John *The Royal Navy: An Illustrated Social History* (Alan Sutton Publishing)

Wragg, D. *The Royal Navy Handbook 1914–1918* (Sutton Publishing)

Admiralty Board Report on the Findings of the Phillips Report 15 September 1944

BR 224/45 Gunnery Pocket Handbook

BR 1661 Handbook for Asdic Set Type 147F

BR 1667 Handbook for Asdic Set Type 147B

BR 1671 Handbook for Asdic Set Type 164

BR 1569 Handbook for DF Outfit FH4

Journal of the Royal Naval Scientific Service Vol 20 No 4 July 1965

Middleton Steering Committee Report on the formation of the Electrical Branch of the Royal Navy January 1946

OU 5245 RN Handbook of Mainguard for Ring Main Protection

Phillips Report on the Torpedo, Anti-submarine, Ordnance and Electrical Branches March 1944
Torpedo School Reports 1880–1922
Tudor Report on the future training of Officers for the Engineering and Electrical Duties of the
    Naval Service 29th July 1921

Live Wire
'HMS Defiance' Vol 1 No 1 Easter 1949
'Once upon a Time' Vol 1 No 3 Christmas 1949
'A Brief History of the Naval electrician' Vol 2 No 1 Easter 1950
'The Old Days and the Old Ways' Vol 2 No 1 Easter 1950
'Electric Signalling without Wires' Vol 2 No 2 Summer 1950
'Electricity and the Gun' Vol 3 No 1 Easter 1951

Naval Electrical Review
'Electrical Engineering in the Royal Navy' Vol 14 No.3 January 1961
Leaving Address by Rear Admiral Sir Kenneth Buckley KBE from the Naval Electrical Review
    Vol 15 No 2 October 1961

Naval Radio Review
'German Guided Missiles' July 1947

Naval Radio and Electrical Review
'The Introduction of AC for Main Power Systems in HM Ships' Vol 2 No. 4 April 1949
'Electrics in the Yangtse Action' Vol 3 No. 2 October 1949
'Remember your batteries! Vol 1 No.1 March 1947
'Genesis Part 1' Vol 8 No.2 October 1954
'Genesis Part 2' Vol 8 No.3 January 1955

END NOTES

1   By the middle of the seventeenth century Royal Navy ships started to be categorised, initially by displacement and then by armament, to reflect the warfare capability they contributed to a Line of Battle.

2   In hindsight, a number of these trials were considered as flawed because the variable quality and preservation state of the iron plating used.

3   The Admiralty has always appeared reluctant to introduce a new form of propulsion system without the backup of the previous technology.

4   The 'Galleass' was a hybrid ship with a combination of three masts and a single bank of oars, briefly featured in the Royal Navy's sixteenth-century order of battle and endured in some Mediterranean fleets until the eighteenth century.

5   One perceived problem with the paddle wheelers was the constraint placed on fitting a sailing rig.

6   Exploding shells had been introduced as early as the seventeenth century but only for use by specialist bomb vessels in the bombardment of shore installations where accuracy was less critical.

7   One of the earliest documented battles between two ironclads occurred in 1862 during the American Civil War. The *USS Monitor* fought the *CSS Virginia* in a stalemate engagement due to the inability of the solid shot ordnance to penetrate the respective hulls.

8   A training correction to compensate for the angular difference in target bearing when observed from the gun direction position and the gun mounting.

9   Sited port and starboard but displaced one forward and one aft of the midships line.

10   This was probably due to the barrel momentum and the embryonic state of electrical control engineering at the time.

11   Major calibre gun ammunition had previous comprised separately loaded charge and shell elements.

12   In 1905, Fisher maintained that of 193 major ships in the Battle Fleet, (excluding destroyers and below), only 63 were of 'such calibre as not to cause the Admiralty grave concern if allowed to wander from the protection of larger ships.'

13   The Fishpond was a somewhat derogatory term given to the circle of Fisher's selected, uniformed associates which included officers dubbed by Fisher as the 'Five best brains in the Navy'.

14   It was perhaps convenient for Churchill that Admiral His Serene Highness Louis of Battenberg, a German prince, was felt by public opinion to be compromised as First Sea Lord following the outbreak of hostilities.

15   The BIR gave birth to the concept of a civilian research arm and over the

years went through several reorganisations before the naval element became the Royal Naval Scientific Service on 8 September 1944.

16   Quote taken from the 1882 Annual Report of the Torpedo School.

17   Captain, later Admiral, Scott was an enthusiastic specialist in the science of gunnery and is credited with developing technology, procedures and practice environments which contributed greatly to the gunnery reputation of the Royal Navy and to the establishment the anti-aircraft defences for London in the First World War.

18   Later Commander-in-Chief of the Grand Fleet at the Battle of Jutland.

19   The surface firing mode was referred to as Low Angle firing until the end of the Second World War. Anti-aircraft firings were referred to as High Angle firings. High Angle firings could be further categorised as long range or close range, with the latter often being referred to as barrage fire, a technique used by smaller calibre weapons, such as pom-poms, to set up a wall of exploding shells at a pre-set close range.

20   The fire control solution comprised the gun elevation and firing bearing which, allowing for the time of flight of the shell, was needed to hit a target at its predicted or future position.

21   Taken from the Fisher Papers entitled 'Naval Necessities' c.1905.

22   The fearnought material used was the same type of flannel material then being used to make stokers' protective trousers and subsequently used in the protective clothing used by ship firefighting parties.

23   Quote taken from 'A Short Treatise on Electricity and the Management of Electric Torpedoes' by Commander J.A. Fisher c.1868.

24   Notwithstanding Fisher's confidence, the torpedo school was still reporting problems with damaged bows and spars in 1881.

25   The Japanese had three battleships and six cruisers sunk or badly damaged by Russian mines during one engagement in the Straits of Tsushima.

26   Nets became the main method of ship defence against the torpedo and deploying nets became a ship's drill throughout the capital ships of the Fleet for the next 30 years.

27   The Brennan torpedo was mechanically driven by reeling in wire from two torpedo mounted spools and using the reaction to drive the weapon through the water. Guidance was obtained by altering the speed of reeling in on one spool and using the speed differential to change the rudder angle. It was not suitable for seagoing service as the controlling winch needed to be fixed in position to maintain guidance control.

28   Admiral Sir George Callaghan did not reap the benefits of his foresight as First Lord of the Admiralty; Winston Churchill, replaced him with Admiral Sir John Jellicoe on 4 August 1914.

29   TNT was expensive to produce and when ammonium nitrate was added to produce Amatol, the explosive properties were enhanced and the cost was reduced.

30   The coherer was invented in 1890 by Edouard Branly and later improved by Sir Oliver Lodge for use as a production device for the detection of radio waves. In principle, it consisted of a glass tube containing metal filings, which in loose form had a very high resistance to electrical current but, in the presence of radio waves, these filings 'cohered' together and formed a conducting path between two output terminals. This path then allowed the passage of a current, thus acting as a detector.

31   Morse code had been invented by this time and was being used to send signals along fixed telegraph lines.

32 The apparatus was later designated as wireless telegraphy, and subsequently radio.

33 Derived from Marconi's demonstrated method of linking transmitters and receivers to operate on a common frequency by the insertion of identical tuning coils in the associated aerial circuits, with each set of coils providing one transmitted 'tune' or frequency, thereby excluding all other untuned systems from receiving any signal.

34 To assist with radar research prior to 1938, frequencies above 30MHz were often referred to as Decimetric Waves (30–3000MHz) or Centimetric Waves (3000–30,000MHz).

35 This responsibility also included underwater communications but not research into hydrophones which remained with the Torpedo School.

36 Following an international treaty, in 1940 the use of spark transmitters was finally banned except for use in ships as an emergency transmitter where it was restricted to 300 watts.

37 Only two transmitters, Types 81 and 83, had a voice capability; they were designed around 1925.

38 The phenomenon whereby HF transmissions are reflected from the ionosphere down to the earth's surface and then back up in the direction of the transmission until the power was fully attenuated. Under good ionospheric conditions this type of propagation could give global communications coverage.

39 The existence of the ionosphere and two fundamental propagation paths, a direct ground wave path following the earth's curvature and an indirect sky wave path reflecting off the ionosphere, and an overall relationship with the radio transmission frequency in use was proven in the 1920s by Messrs Appleton and Barnett.

40 A magnetic field induced into steel ships by the earth's magnetic field as a result of the manufacturing process during the build period.

41 This was considered the top end of the MF band as classified at the time.

42 Reflections by B.W. Lythall, Chief Scientist Royal Navy, on Cecil Horton, father of British radar, taken from *The Development of Radar Equipments for the Royal Navy 1935–45* by F.A. Kingsley.

43 The acronym RDF is attributable to Sir Robert Watson-Watt, and it was taken to stand for either Radio Direction Finding or Ranging and Direction Finding. The latter was more descriptive of the system functionality, but the former, believed to be the official version, was encouraged in order to maintain secrecy and confuse those who did not have security clearance.

44 In 1943, the term RDF was dropped by the Royal Navy when the Allies adopted the acronym 'radar' which stood for Radio Detection and Ranging.

45 1MHz is equivalent to 1 megacycle per second in the terminology of the time.

46 First Sea Lord 1910–1911.

47 The Submarine Committee was replaced by the Anti-submarine Division of the Admiralty on 18 December 1916.

48 In January 1918, the BIR was reorganised and replaced by the Department of Experiment and Research (DER), which reported to the Admiralty Board through the Third Sea Lord. DER was itself replaced in January 1919 by the Department of Scientific Research and Experiment (SRE), also headed by the Third Sea Lord.

49 Lieutenant Harty was later Sir Hamilton Harty, the conductor of the Halle Orchestra.

50 A combined fleet of cruisers, destroyers and submarines charged with

protecting the northern approaches to the English Channel.

51   Fessenden invented this mechanical oscillator which was eventually used for
      communications, echo detection and depth sounding applications.

52   British magnetic mines were available before the end of the First World
      War but were not used operationally because of the fear that a mine might
      fall into enemy hands, and the Royal Navy did not have a minesweeping
      countermeasure.

53   Allegedly, this acronym was first used by Winston Churchill after he referred
      to the top secret acoustic programme as 'asdic' in Parliament in 1939. In
      response to a question from the etymological department of the *Oxford
      English Dictionary*, the term was explained as standing for Allied Submarine
      Detection Investigation Committee. Although there are no records of there
      actually being such a committee, this alternative definition is still widely used
      in asdic folklore.

54   The P Class patrol boats were unassuming, shallow draft ships built under
      the Emergency War Programme of 1917 for anti-submarine operations.
      They were designed to lure U-boats into carrying out an attack and then
      counterattacking at unexpected speed using depth charges or a purpose built
      ramming bow.

55   The arc of water ensonified by the transducer.

56   The bathythermograph – used for measuring the temperature variation in
      different water depths – was not invented until 1937 when it became an
      important tool for predicting asdic performance.

57   The bolometer was invented in 1878 by Samuel Langley. In the case of
      this transmission system, the principle of the Wheatstone Bridge was used
      whereby variation of resistance in the bridge was obtained by allowing the
      heat of a lamp to play on high temperature coefficient resistive material
      through a slit in a disc which was controlled by the magnetic compass card.

58   Somewhat confusingly, the term 'branch' has been used by the Royal Navy
      to describe the grouping of military or non-military officers and, as manning
      structures evolved and became more complicated, specialist and sub-specialist
      groups.

59   Around 1800 the Navy Board reported to the Admiralty Board for certain
      administrative functions including warrant officer appointments.

60   In peacetime, many ships were put into 'ordinary' care or Reserve. The
      commissioned officers and many crew were discharged to shore and the
      standing officers supervised a skeleton crew for maintenance purposes only.

61   A circle of gold lace which still exists and sits above the ring(s) of gold
      distinction lace on the sleeve denoting the ranks of the wearer.

62   The resistance to this concept was so great that it was never introduced and
      the Royal Marines remained a separate entity.

63   A perceived more public humiliation was that because Military List officers
      were listed separately and appeared before the Civil List officers in the Navy
      List, thus junior engineers appeared before their seniors who were not
      Selborne-Fisher qualified.

64   This was the first time that no separate reference is made to the pre Selborne-
      Fisher engineers in the Uniform Regulations for branch distinguishing
      uniform features. Previously they had appeared as a separate branch with
      mention made of the purple distinction cloth but wearing of the executive
      curl specifically excluded.

65   Probably the first official use of word 'executive' to define a branch, rather
      than a specialisation.

66 Perhaps indicative of the decline in sail, *Algiers* was never actually fitted out for sea and spent most of her naval service as a hulk before being scrapped in 1904.

67 The Naval Discipline Act was first implemented in 1860, partly in response to public outcries against flogging. It defined a Captain's powers of summary punishment in an attempt to standardise ship board methods of dealing with offenders.

68 It was not until towards the end of the nineteenth century – following the introduction of the ship's Low Power Switchboard – that generators started to replace batteries as a source of power for firing circuits.

69 Wordingham eventually became the President of the Institution of Electrical Engineers in 1917.

70 A note made by *Defiance* in the 1901 Torpedo School Report indicated that many recruits were failing the course through 'being deficient in lathe work'.

71 No commissioned officers from warrant rank would have been promoted by the start of the First World War.

72 In the 1930s the Engineer Branch again took over responsibility for the electrical machinery in its spaces and the ERA started to receive electrical training once again, but only until the Electrical Branch was formed in 1946.

73 At the same time the boy ordnance artificer was introduced to support the newly founded ordnance artificer.

74 The RNVR was formed following the 1903 Naval Forces Act to provide officer and rating support for the Royal Navy, both ashore and afloat. It became known as the 'Wavy Navy' because the officers wore zig zag gold distinction lace.

75 At this time, under the principles of the Selborne-Fisher Officer Training Scheme, midshipmen were given the option to volunteer for their Branch of choice on completion of a common training period including time at sea.

76 The use of two or more transmitter aerials spaced an integer number of wavelengths apart to ensure different propagation paths of the same signal and produce uncorrelated fading of that signal. Similarly spaced receiver aerials could then receive the same signal from which the best could be selected, or the total be combined, to achieve an output which was comprehensible.

77 FM had been around since the 1920s but was not an internationally adopted standard and, even as late as 1945, the Royal Navy was not convinced that there were clear advantages over AM. Reduced size and weight was an aspect which made the technology suitable for aircraft but the RAF did not opt for FM until after the war.

78 In August 1941 the Experimental Department of the Royal Navy Signal School had been re-designated as the Admiralty Signal Establishment.

79 Single aerial working, using a diode switching device, was eventually achieved in March 1941 following trials in HMS *Hood* and this configuration was then used as the basis for all further radar system design.

80 Continuous rotation was not yet technically possible in any radar set but this was not considered a serious limitation until the invention of the Plan Position Indicator which gave a 360° display about the ship. A version of the Type 281 with continuous rotation was eventually produced in 1945.

81 The Royal Navy supplemented the Chain Home coverage on a more permanent basis by installing a Type 79Z radar at Fort Wallington near Fareham, Hampshire with another Type 79Z, installed at Eastney for instructional purposes, as emergency backup.

82 In fact these results were over optimistic due to the trials being carried out under conditions of 'Anomalous Propagation'. This was an atmospheric

phenomenon unknown at the time but later discovered to be the channelling of higher frequency transmissions near the sea surface giving extended detection ranges in a manner similar to the mirage phenomenon.

83   Estimate taken from the list of naval radars in *Confidential Book CB 4497, Simple Guide to Naval Radar circa 1949.*

84   This term refers to observing the disposition of the forces shown on the display and assessing the tactical implications of that disposition.

85   The use of 440 volts AC for domestics, including lighting, was deemed to be a safety problem but this was overcome by accessible domestic equipment, including lighting, being powered by 115 volts AC transformed down from the 440 volt AC used to supply the ship's ring main.

86   At this time, the calculation would not have predicted the massive savings in power requirements which would accompany technology advances in the electronics field.

87   The greatest depth confirmed by the Royal Navy for U-boat operation during the war was 780 feet.

88   The asdic transmission was at a frequency which could be heard by the human ear. Hence the 'pinging' could be heard.

89   Two heavier mortar systems were also under consideration but these were dismissed for parochial divisions between both uniformed and scientific communities.

90   At one time under Captain John Walker's command and part of the 2nd Escort Squadron.

91   Taken from post-war analysis which showed comparable figures of 6 per cent for depth charges and 20 per cent for Hedgehog.

92   High explosive shells were used primarily to incapacitate. The alternative was the armour piecing shell which was designed to explode after penetrating any armour plating.

93   Also referred to as the Velocity Trigger Fuze and later as the Variable Time Fuze.

94   Designated as such in King's Regulations and Admiralty Instructions Article 1244.

95   The emerald green stripe was not just a 'radar' designator as it was also used by all specialist officers brought into the service during the war.

96   Shore based numbers were supplemented by female recruits brought into the WRNS Torpedo Branch with many reaching LTO status before the end of the war.

97   By this time in the war, *Vernon* had been evacuated away from Portsmouth to escape the Blitz and been relocated at Roedean Girls School in Brighton, where it was designated as *Vernon*(R).

98   Square rig was the standard rig for leading rate.

99   The evaluation of the operational environment, setting of operational policy and specification of a Naval Staff Requirement for an operational capability.

100 In early 1941 matters were much improved by the arrival of some 20 Canadian RNVR officers with radio physics degrees, who were loaned to the Royal Navy courtesy of the Canadian Government for radar duties because of the shortage of suitably qualified British graduates.

101 Time in the service spent over the age of seventeen years six months, the only time which counted towards the serving of an engagement.

102 'Jackie' Fisher had introduced the mechanician rate in 1905 to improve the career prospects of the stoker by providing a route to warrant rank and to support the ERA by undertaking some of their watchkeeping duties. By 1914 there were 72 warrant mechanicians in the service.

103 The day to day nomenclature used for the artificer was to be either electrical artificer or radio artificer. It was not until the Series Scheme was introduced in 1947 that the term radio electrical artificer formally came into use.

104 It should be remembered that while these matters were being put forward, the amalgamation of the Torpedo and AS Branches was also being similarly discussed.

105 As classified in Middleton's Report, Grade 2 work was skilled maintenance, fault finding, minor overhauls and repairs, while Grade 3 work comprised semi-skilled maintenance and simple repairs.

106 This point had been identified as a principle issue of the Middleton's Report on the Electrical Branch.

107 The Chatham Torpedo School had initially been established at Sheerness in 1905 when the *Ariadne*, a hulk, was moved from Portsmouth, where it had been part of *Vernon*. *Ariadne* was renamed *Actaeon* and when the torpedo school moved ashore it was set up in *Actaeon* Building. Unlike the other torpedo schools, the Chatham School was not commissioned as HMS *Actaeon*, probably because the name had been reallocated to an operational ship.

108 The EA(R) rate was re-designated to REA on 1 January 1948.

109 Within the Civil Service, the Royal Naval Electrical Service retained the green stripe but they were not normally considered as an active duty branch.

110 For some reason the *Naval Radio and Electrical Review* of January 1949 quotes the derivation as a contraction of 'transfer resistor' but the internal minutes of a Bell Laboratory meeting held to discuss the name of the new component gives the true origination.

111 Prime examples of systems using complex analogue fire control computers were MRS3, AFCB10 and MCS10.

112 Taken from *Naval Electronic Warfare* by Dr David G. Kiely, Volume 5 of Brassy's Seapower series.

113 These wavelengths covered I band (*c.*9MHz) and S band (*c.*3Mhz) radar transmissions.

114 Under the current international standard, the UHF band covers 300–3000MHz.

115 WEE seamen were seamen who worked for the WEEO in maintenance and quarters roles. Many of these men were volunteers for permanent duty with the WEE department with a considerable number applying to transfer to the WEE Branch after enjoying the technical experience.

116 Arguably, this decline was due to the introduction of the principal warfare officer concept which had displaced the gunnery and TAS specialist officers who were given in depth training and experience of their armament stores.

117 At the time, the embracing action priorities for Royal Navy ships were to fight, to move and to float with the decision to move from one to another state being made by the command.

118 CACS2 had been intended as an upgrade for the Type 42 destroyer and CACS3 for the Type 43 destroyer. Both programmes were cancelled.

119 The Royal Navy's annual publication, sponsored by the Navy Board and containing a wide ranging view of the operations in that year and perspective view of the future.

120 By 2009 this view was beginning to change with the swing back to apprenticeships being hailed as a return to work based vocational training which had been lost during the Labour government's political drive to widen the application of degree qualifications into vocational areas and introducing generic theoretical knowledge to justify a three-year degree at the expense of gaining practical skills.

121   In charge of tri-Service recruitment of the scientists, engineers and uniformed personnel needed for the Second World War radar effort.
122   Someone who has crossed the equator many times.
123   *Osprey* was the name for the Asdic Training School, Portland, Dorset.
124   A class of destroyers mass produced for the US Navy in the First World War which played a vital role during the early years of the Second World War.
125   Jim Menter returned to Cambridge after the war and later became Sir James Menter, a distinguished scientist and industrialist.
126   A draft to sea without notice.
127   RDF terminology in the RAF and Royal Navy's developed independently during the war, 'blips' being the RAF's term for the Royal Navy's 'echo'. A committee sat down after the war to rationalise the vocabulary and one outcome was the Royal Navy's adoption of the term 'radar'.
128   In Derek Howse's book *Radar at Sea* this story is confirmed by Sub Lieutenant A.H.G. Butler RNVR who was serving in HMS *Westminster* when the three ball fictional Type 298 was fitted. One enquiry signal apparently came from Admiral Fraser in HMS *King George V* who signalled C-in-C Rosyth 'Understand you have Type 298 fitted in one of your Escort Force. Would much appreciate full particulars.'
129   Nothing changes: In 1982 HMS *Newcastle* off the Falklands had both 909s down. The aft radar 909 matching stub was successfully replaced by an araldite stub of 'local' manufacture. The outstanding problem was how to clear the Opdef without having an official spare.

INDEX

(Collingwood Magazine Archives)